Publishing and Using Cultural Heritage Linked Data on the Semantic Web

Synthesis Lectures on Semantic Web: Theory and Technology

Editors
James Hendler, *Rensselaer Polytechnic Institute*
Ying Ding, *Indiana University*

Synthesis Lectures on the Semantic Web: Theory and Application is edited by James Hendler of Rensselaer Polytechnic Institute. Whether you call it the Semantic Web, Linked Data, or Web 3.0, a new generation of Web technologies is offering major advances in the evolution of the World Wide Web. As the first generation of this technology transitions out of the laboratory, new research is exploring how the growing Web of Data will change our world. While topics such as ontology-building and logics remain vital, new areas such as the use of semantics in Web search, the linking and use of open data on the Web, and future applications that will be supported by these technologies are becoming important research areas in their own right. Whether they be scientists, engineers or practitioners, Web users increasingly need to understand not just the new technologies of the Semantic Web, but to understand the principles by which those technologies work, and the best practices for assembling systems that integrate the different languages, resources, and functionalities that will be important in keeping the Web the rapidly expanding, and constantly changing, information space that has changed our lives.

Topics to be included:

- Semantic Web Principles from linked-data to ontology design

- Key Semantic Web technologies and algorithms

- Semantic Search and language technologies

- The Emerging "Web of Data" and its use in industry, government and university applications

- Trust, Social networking and collaboration technologies for the Semantic Web

- The economics of Semantic Web application adoption and use

- Publishing and Science on the Semantic Web

- Semantic Web in health care and life sciences

Publishing and Using Cultural Heritage Linked Data on the Semantic Web
Eero Hyvönen
2012

VIVO: A Semantic Approach to Scholarly Networking and Discovery
Katy Börner, Mike Conlon, Jon Corson-Rikert, and Ying Ding
2012

Linked Data: Evolving the Web into a Global Data Space
Tom Heath and Christian Bizer
2011

Publishing and Using Cultural Heritage Linked Data on the Semantic Web

Eero Hyvönen

ISBN: 978-3-031-79437-7   paperback
ISBN: 978-3-031-79438-4   ebook

DOI 10.1007/978-3-031-79438-4

A Publication in the Springer series
SYNTHESIS LECTURES ON SEMANTIC WEB: THEORY AND TECHNOLOGY

Lecture #3
Series Editors: James Hendler, *Rensselaer Polytechnic Institute*
                Ying Ding, *Indiana University*
Synthesis Lectures on Semantic Web: Theory and Technology
ISSN pending.

Publishing and Using Cultural Heritage Linked Data on the Semantic Web

Eero Hyvönen
Aalto University

SYNTHESIS LECTURES ON SEMANTIC WEB: THEORY AND TECHNOLOGY
#3

ABSTRACT

Cultural Heritage (CH) data is syntactically and semantically heterogeneous, multilingual, semantically rich, and highly interlinked. It is produced in a distributed, open fashion by museums, libraries, archives, and media organizations, as well as individual persons. Managing publication of such richness and variety of content on the Web, and at the same time supporting distributed, interoperable content creation processes, poses challenges where traditional publication approaches need to be re-thought. Application of the principles and technologies of Linked Data and the Semantic Web is a new, promising approach to address these problems. This development is leading to the creation of large national and international CH portals, such as Europeana, to large open data repositories, such as the Linked Open Data Cloud, and massive publications of linked library data in the U.S., Europe, and Asia. Cultural Heritage has become one of the most successful application domains of Linked Data and Semantic Web technologies.

This book gives an overview on why, when, and how Linked (Open) Data and Semantic Web technologies can be employed in practice in publishing CH collections and other content on the Web. The text first motivates and presents a general semantic portal model and publishing framework as a solution approach to distributed semantic content creation, based on an ontology infrastructure. On the Semantic Web, such an infrastructure includes shared metadata models, ontologies, and logical reasoning, and is supported by shared ontology and other Web services alleviating the use of the new technology and linked data in legacy cataloging systems. The goal of all this is to provide layman users and researchers with new, more intelligent and usable Web applications that can be utilized by other Web applications, too, via well-defined Application Programming Interfaces (API). At the same time, it is possible to provide publishing organizations with more cost-efficient solutions for content creation and publication.

This book is targeted to computer scientists, museum curators, librarians, archivists, and other CH professionals interested in Linked Data and CH applications on the Semantic Web. The text is focused on practice and applications, making it suitable to students, researchers, and practitioners developing Web services and applications of CH, as well as to CH managers willing to understand the technical issues and challenges involved in linked data publication.

KEYWORDS

Semantic Web, linked data, cultural heritage, portal, metadata, ontologies, logic rules, information retrieval, semantic search, recommender system

Contents

Preface

Publishing Cultural Heritage (CH) collections and other content on the Web has become one of the most successful application domains of Semantic Web and Linked Data technologies. After a period of technical research and prototype development, boosted by the W3C Semantic Web Activity kick-off in 2001 and the Linked (Open) Data movement later on, major national and international CH institutions and collaboration networks have now started to publish their data using Linked Data principles and Semantic Web technologies.

This work is highly interdisciplinary, involving domain expertise of museum curators, librarians, archivists, and researchers of cultural heritage, as well as technical expertise of computer scientists and Web designers. Applying a new technology in the rapidly evolving Web environment is challenging not only for non-technical personnel in CH institutions, but also for computer scientists themselves.

This book aims at fostering the application of Linked Data and Semantic Web technologies in the CH domain by providing an overview of this fascinating application domain of semantic computing. My own work in this field started in 2001 after the W3C Semantic Web Activity launch by establishing the Semantic Computing Research Group (SeCo) focusing on this field. We first developed a semantic photograph search and recommender system for a university museum, followed by semantic portal prototypes for publishing heterogeneous collections of different kinds, including artifacts in cultural history museums, historical events, folklore, maps, fiction literature, and natural history museum data. This book reflects experiences gained during this work.

From the very beginning in 2002, after developing our first ontologies and transforming the first collection databases into RDF, it became clear that the possibility of reusing existing data, metadata models, and ontologies, and linking it all together in an interoperable way, will be a central benefit of Semantic Web applications. W3C recommendations, such as RDF(S), SKOS, SPARQL, and OWL are the corner stones for facilitating cross-domain, domain-independent interoperability, but this is not enough. We also need domain-dependent metadata-models and domain ontologies based on the generic semantic principles, as well as domain specific datasets. From a practical viewpoint, we also need ontology services so that the shared resources can be published and used in legacy and other application systems in a cost-efficient way. In short, a Semantic Web *content infrastructure* needs to be built in a similar vein as railroad, telephone, and other communication networks were created during earlier technological breakthroughs.

Creating a Semantic Web infrastructure, as well as content for it, requires collaboration between content providers. Co-operation is needed not only for sharing data through joint portals such as Europeana, but also for developing shared metadata models and ontologies used in representing the contents in an interoperable way. Publishing CH content is becoming a game of cross-domain networking where the traditional boundaries of memory organizations based on content types are breaking down. From a user's viewpoint, the focus is on data, knowledge, and experience, be it based on a book in a library, an artifact in a cultural history museum, a story in an archive, a painting in an art gallery, a photograph taken by a fellow citizen, or a piece of music on a record.

During these years my faith in Semantic Web and Linked Data has become strong even if there are great challenges ahead, too. This is a truly promising way for providing richer content to users through more intelligent and usable interfaces, and at the same time for facilitating memory organization with better tools for collaborative, open content publishing on the Semantic Web.

Eero Hyvönen
October 2012

Acknowledgments

Thanks to the series editors Jim Hendler and Ying Ding for the invitation to write this book, and to Mike Morgan for making the publication possible.

The book's contents are based on collaboration with various students, researchers, and visitors in the Semantic Computing Research Group (SeCo) at the Aalto University and University of Helsinki in different times including (in alphabetical order) Matias Frosterus, Harri Hämäläinen, Tomi Kauppinen, Suvi Kettula, Heini Kuittinen, Jussi Kurki, Nina Laurenne, Aleksi Lindblad, Thea Lindquist, Glauco Mantegari, Eetu Mäkelä, Panu Paakkarinen, Tuomas Palonen, Sini Pessala, Tuukka Ruotsalo, Samppa Saarela, Katri Seppälä, Osma Suominen, Jouni Tuominen, Juha Törnroos, Mika Wahlroos, Mark van Assem, and Kim Viljanen.

Ying Ding, Stefan Gradmann, Patrick Leboeuf, Glauco Mantegari, Katri Seppälä, and Regine Stein made fruitful comments to earlier versions of this manuscript. Special thanks to Jouni Tuominen for several comments, suggestions, and help in proofreading the text. C.L. Tondo's help was invaluable in finalizing the text and layout.

Fruitful collaboration with several museums, libraries, archives, and media organizations in Finland is acknowledged, including (in alphabetical order) Agricola.fi network of historians, Antikvaria Museum Group, Espoo City Museum, Finnish Agriculture Museum, Finnish Broadcasting Company YLE, Finnish Literature Society, Finnish Museum of Photography, Finnish Museum Association, Finnish National Gallery, Finnish Public Libraries (Libraries.fi), Helsinki City Library, Helsinki University Library, Helsinki University Museum, Lahti City Museum, National Board of Antiquities, National Library of Finland, and Suomenlinna Sea Fortress.

The National Funding Agency for Technology and Innovation (Tekes)[1] and consortia of tens of public organizations and companies have supported several research projects of SeCo related to CH, such as Intelligent Catalogs (2002–2004), FinnONTO[2] (2003–2012), Semantic Ubiquitous Services (2009–2012)[3], and Linked Data Finland[4] (2012–). The Finnish Cultural Foundation[5] has supported our research on the CultureSampo system, too.

Thanks to SmartMuseum EU project[6] for funding and collaboration, to European Institute of Technology (EIT) Project EventMAP, as well as to the Network for Digital Methods in the Arts and Humanities (NeDiMAH) (European Science Foundation). Joint work with the University of Colorado regarding war history and linked data is acknowledged. Thanks to collaborations with the

---

[1]http://www.tekes.fi/en/
[2]http://www.seco.tkk.fi/projects/finnonto/
[3]http://www.seco.tkk.fi/projects/subi/
[4]http://www.seco.tkk.fi/projects/ldf/
[5]http://www.skr.fi/
[6]http://www.smartmuseum.eu/

Continuous Access to Cultural Heritage (CATCH) initiative and colleagues at the VU University and other universities in the Netherlands.

Eero Hyvönen
October 2012

CHAPTER 1

Cultural Heritage on the Semantic Web

Cultural Heritage (CH) refers to the legacy of physical objects, environment, traditions, and knowledge of a society that are inherited from the past, maintained and developed further in the present, and preserved (conserved) for the benefit of future generations[1]. This chapter first characterizes the notion of CH and identifies specific challenges encountered when publishing CH contents, especially collection data, on the Web. After this, Semantic Web and Linked Data technologies are introduced as a novel, promising approach to address the problems. The chapter ends with an overview of the book content.

1.1 CHARACTERIZING CULTURAL HERITAGE

CH can divided into three subareas.

1. **Tangible cultural heritage** consists of concrete cultural objects, such as artifacts, works of art, buildings, and books.

2. **Intangible cultural heritage** includes phenomena such as traditions, language, handicraft skills, folklore, and knowledge.

3. **Natural cultural heritage** consists of culturally significant landscapes, biodiversity, and geodiversity.

The key players in preserving CH are *memory organizations* that include libraries, archives, and museums of different kinds specializing in particular areas of CH, such as art museums, archaeological museums, botanical museums and gardens, cultural history museums, medical collections, science museums, theater history museums, geological and mineralogical museums, and zoology museums. Also media organizations often preserve CH materials, especially more recent ones. There are also lots of CH materials maintained by cultural associations of various kinds and individual persons. Tangible CH objects are stored with attached metadata, intangible heritage is documented using textual descriptions, photographs, interviews, and videos, and there are natural history and other museums specializing in storing traces and knowledge of natural history, geology, and environment.

---

[1]In this book, the ambiguous term "culture" is used to refer to the "the ideas, customs, skills, arts, etc. of a people or group, that are transferred, communicated, or passed along, as in or to succeeding generations" (Webster's New World Dictionary).

The Web has become an increasingly important medium for publishing CH contents of different kinds. For example, libraries and archives are online with their collections, museums show their collections through collection browsers, and documentation of intangible heritage is available as audio and video recordings and as interactive hypertext applications, even as games. There are large national and multi-national CH portal projects active in harvesting and publishing content from different sources via centralized services.

For the layman end-user, such systems provide a single access point to massive heterogeneous collections and an authoritative source of information. In contrast to traditional physical exhibitions, Web services are open all the time, can be accessed without physical presence at an exhibition, the number of exhibits on the Web is not limited by the physical space available, and the exhibits can be linked and accessed flexibly using different strategies, not only the one used in the physical exhibition. Of course, the Web cannot replace the physical experience of visiting a museum or an exhibition in reality but provides a complementary alternative for accessing collection data virtually at any time and from any place.

For researchers in the humanities, availability of CH data in massive amounts in digital machine processable form has opened up a new research paradigm called Digital Humanities.

1.2   INFORMATION PORTALS FOR CULTURAL HERITAGE

There are several kind of CH publications on the Web. First, there is a large variety of well-curated systems that have been hand-crafted for a specific purpose with a focused closed theme, dataset, and interfaces. Such systems are often implemented using tools such as Adobe Flash with a beautiful game-like appearance. For example, the Lewis and Clark Expedition (1803–1806) is documented on the Web in great detail by several applications. The portal in Figure 1.1[2] provides the end-user with several thematic perspectives to the journey by selecting the buttons on the left, such as "overview," "American nation," "geography," "journal excerpts," "natural history," and "technology used." Such systems may also be available on CD/DVD as stand-alone applications.

On the other end of the spectrum, there are collection search services and browsers providing access to large open collection databases whose content is not thematically focused, and curated access paths and interfaces may be missing. In return, large collection databases originating possibly from several institutions can be accessed. For example, a variety of Australian CH collections can be accessed using the Collections Australia Network system[3]. Similar federated portals for searching and browsing collections can be found in many countries and internationally. A flagship application here is Europeana[4], based on millions of collection objects originating from memory organizations all over Europe. For example, in Figure 1.2 the user has typed in the keyword "chair" in the search field of Europeana and the system has found various chairs in participating collections. The search can be refined further by selecting additional filters on the facets on the left, such as "media type,"

---

[2]http://lewis-clark.org/
[3]http://www.collectionsaustralia.net/
[4]http://www.europeana.eu/

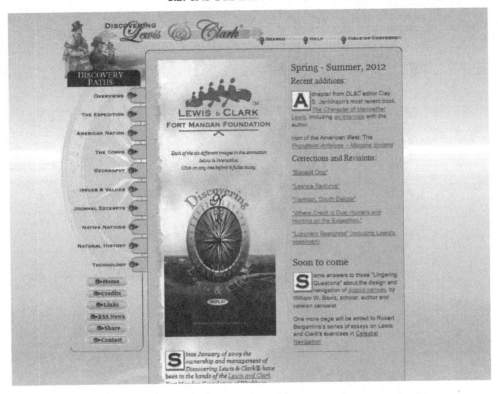

Figure 1.1: A portal exhibiting versatile content related to the Lewis and Clark expedition (1803–1806) in the U.S. from different perspectives. (Fort Mandan Foundation, North Dakota)

"language," "date," "country," and whether content contributed by users should be included or not. Another portal example, harvesting library data, is WorldCat[5] that contains metadata (without the primary sources) of about 1.5 billion books, DVDs, CDs, and articles in the participating libraries. The World Digital Library[6] is yet another international portal, operated by UNESCO and the United States Library of Congress, that makes available, free of charge, significant multilingual primary materials, such as manuscripts, maps, rare books, musical scores, recordings, films, prints, photographs, and architectural drawings.

In this book, the main focus is on *information portal systems* of the latter kind: CH portals based on large heterogeneous collection datasets are considered, where organizing the contents by hand into a focused thematic application with application-specific visualizations and interfaces is not usually feasible. Such shared publication portals facilitate exchange of knowledge for CH researchers, librarians, and archivists. For the contributing memory organizations, such systems are

---

[5]http://www.worldcat.org/
[6]http://www.wdl.org/

Figure 1.2: Faceted search in Europeana portal exhibiting chairs from different European collections.

an opportunity to reach out to wider audiences on the Web with new ways of interaction, and to collaborate with other organizations. From a societal perspective, publishing CH on the Web stimulates cultural tourism, creative economy, and enhances friendly relationships and unity between parties and nations involved in such initiatives.

1.3   CHALLENGES OF CULTURAL HERITAGE DATA

CH collection data has many specific characteristic features, such as the following.

- **Multi-format**. The contents are presented in various forms, such as text documents, images, audio tracks, videos, collection items, and learning objects.

- **Multi-topical**. The contents concern various topics, such as art, history, artifacts, and traditions.

- **Multi-lingual**. The content is available in different languages.

- **Multi-cultural**. The content is related and interpreted in terms of different cultures, such as religions or national traditions in the West and East.

- **Multi-targeted**. The contents are often targeted to both laymen and experts, young and old.

As a result, a fundamental problem area in dealing with CH data is to make the content mutually *interoperable*, so that it can be searched, linked, and presented in a harmonized way across the boundaries of the datasets and data silos. The problem occurs on a syntactic level, e.g., when harmonizing different character sets, data formats, notations, and collection records used in different collections. Even more importantly, there is the problem of *semantic interoperability*: different metadata formats may be interpreted differently, data is encoded at different levels of precision, vocabularies and gazetteers used in describing the content are different, and so on. The Semantic Web standards[7] and best practices, especially those advocated by the World Wide Web Consortium (W3C)[8], provide a shared basis on which interoperable Web systems can be built in a well-defined manner. The new technologies are of course no panacea for all problems but rather a tool set by which the hard issues can be tackled arguably more effectively than before.

A major reason for interoperability problems in CH content publishing is the *multi-organizational* nature in which CH content is collected, maintained, and published. The content is provided by different museums, libraries, and archives with their own established standards and best practices, by media organizations, cultural associations, and individual citizens in a Web 2.0 fashion. The success of the WWW is very much due to its simple distributed many-to-many publishing paradigm that has few restrictions and shared standards, with the HTML mark-up language combined with the HTTP protocol and the idea of URL addressing as core technologies. However, things get more complicated on the Semantic Web, where content is not published only for human users in HTML form but also as data for machines to use. An additional standard base is needed for the Web of Data. In application domains such as CH more coordinated collaboration is needed between CH publishers and the technical WWW developer community than before.

1.4   PROMISES OF THE SEMANTIC WEB

Semantic Web technologies[9] [34] (SW) are a promising new approach for addressing the problems of publishing CH content on the Web. The term "semantic" here refers to *Semantics*, a discipline studying relations between *signifiers*, such as words, phrases, signs, and symbols, and what they stand for, i.e., *denotata*. In Computer Science semantics refers to the formal meaning and interpretation (declarative or procedural) that has been given to syntactic structures, such as programming languages or symbolic data structures.

---

[7]Called "recommendations" by the W3C.
[8]http://www.w3.org/
[9]http://www.w3.org/standards/semanticweb/

The Semantic Web can be seen as a new layer of *metadata* being build inside the Web. According to the traditional definition, metadata is data about data. For example, a metadata record of a book (data) may tell its title, author, subject, and publishing year. However, the term "metadata" is used more widely in the Semantic Web context as a synonym for machine processable or interpretable data. The key idea is that syntactic metadata structures make Web content "understandable" to the machines, based on shared semantic specifications founded on formal logic. This makes it possible to create more interoperable and intelligent Web services. A computer that cannot interpret the data it is dealing with is like a telephone just passing information, and cannot be very helpful in more complicated information processing tasks dealing with the meanings of the contents.

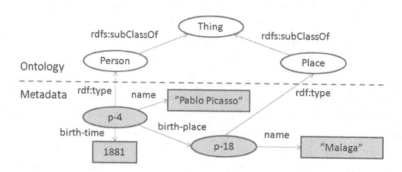

Figure 1.3: The data model of RDF is a directed labeled graph.

The methodology for representing metadata and ontological concepts[10] on the Web is based on a simple data model: a directed labeled graph, i.e., a *semantic net*. For example, Figure 1.3 depicts an RDF graph telling on a metadata level that the identity *p-4* is an individual of the class *Person* (denoted by the arc *rdf:type*) with name "Pablo Picasso" born in 1881 at an instance *p-18* of the class *Place* whose name is "Malaga." In the RDF graph, classes such as places and persons are represented as subclasses (arc *rdfs:subClassOf*) of the class *Thing* on an ontology level, while the individuals of the classes are considered metadata. Both metadata and ontologies are represented uniformly in the same graph. In the figure, identities that may have properties, i.e., may have out-going arcs, are depicted as ovals while literal terminal atomic values without further properties (here strings and numbers) as rectangular boxes.

The figure illustrates that actually there are several levels of descriptions needed on the Semantic Web.

1. **Real world**. On the bottom, there is the real world, i.e., the domain of discourse, such as persons, artifacts, and places.

---

[10]The notion of "concept" is a complex philosophical notion referring to a general idea or something conceived in the mind. On the Semantic Web, the term "concept" is used for any entity on the Web or outside of it with an identity specified by a URI.

2. **Data level.** Then there is the data level, since real world items have to be represented as data of some kind in a computer. For example, images and documents are data as well as a URI reference to a person.

3. **Metadata level.** After data, there is metadata about the data, e.g., records in a collection database about images, persons, or artifacts.

4. **Ontology level.** Next, ontology level defines the generic classes and properties used in describing a domain, i.e., the vocabularies in terms of which the metadata is represented. The metadata schema used in cataloging and controlled vocabularies of subject headings are part of this level. For example, in Figure 1.3 persons are described in terms of their name, birth time, and birth place, and instantiated from the classes defined on the ontology level. The same ontologies can be used for representing collection metadata of a similar domain area in different memory organizations (e.g., books in libraries).

5. **Metaontology level.** Finally, there are the general cross-domain modeling principles of ontologies that are domain-independent. For example, the notions of subclass-of relation and class are generic and not restricted to a particular domain. Such generic principles are specified by the Semantic Web standards, such as RDF(S) and OWL, and facilitate cross-domain interoperability of contents.

On a global WWW scale, the Semantic Web forms a *Giant Global Graph* (GGG) of connected data resources. The GGG can be used and browsed in ways analogous to the WWW, but while the WWW links associated Web pages with each other for human use, the GGG links associated underlying concepts and data resources together. For example, the GGG may tell that ducks are birds, and that Donald is an instance of a duck (and therefore a bird) while the related WWW pages may constitute a comics book about Donald Duck.

A key idea of linked data is that the different parts of the GGG can come from different data sources. For example, in Figure 1.3 metadata about persons, such as Pablo Picasso, may come from an authority database, information about places, such as Malaga, may be provided by a land survey organization, and the class ontology can be based on an existing keyword thesaurus in use in a library. Different data sources are illustrated in the figure by different colors/densities.

Based on harmonized RDF-based representations of data, more "intelligent" Web applications can be built and with less effort. From a technical application perspective, Semantic Web technologies have many promising features:

- *More accurate content descriptions.* The technology is based on globally unique Universal Resource Identifiers (URI), which makes it possible to refer to meanings more accurately than using literal expressions. For example, person and place names can be disambiguated: there are lots of "John Smiths" around, "Paris" can be found in France, Texas, and in many other places, and the names can have different transliterations in different language systems. In libraries, the notion of, e.g., Shakespeare's play "Hamlet" can refer to the abstract story, its manifestation as a text or a video of the play, different translations of it, variants of the story, editions of these, and finally individual books or DVDs on the library shelves. Modeling such semantic distinctions can be done using novel "ontology-based" CH standards to be presented in this book.

- *Interoperability.* Semantic Web technologies provide a novel approach to creating interoperable linked data.

- *Simple data model for aggregation.* Two (interoperable) RDF graphs can be joined together technically in a trivial way by simply making the union of them (i.e., the corresponding triple sets).

- *Data aggregation by linked data.* By combining data sources in an interoperable way, data from one source can be enriched with additional linked data from another source. A notable international initiative toward this goal is Linked Data[11] [53], where open datasets such as Wikipedia/DBpedia[12] and Freebase[13] for common knowledge, GeoNames[14] for millions of place names, or Gutenberg project[15] for over 40,000 free ebooks are described in terms of Semantic Web standards and interlinked with each other.

- *Semantic Web services.* Semantic linked data is published not only as passive datasets, but as operational services than can be utilized by legacy and other CH applications via open and generic Application Programming Interfaces (API). By utilizing shared ready-made services, application programmers can re-use work done by others, and save their own programming effort and resources. This idea can be paralleled with Google and Yahoo! Maps that provide map services on a global basis to applications via easy-to-use APIs for mash-up development.

Publishing CH on the Web is not only a technical challenge; issues of trustworthiness of content, copyrights, and licensing are also of concern. Much of CH content is protected by copyright, and there are also other reasons why organizations cannot publish their data openly, e.g., issues of personal privacy. However, based on the ideas of Linked Open Data, the WWW world is clearly taking steps toward publishing open data and free of charge when feasible. The idea is that CH content should be maximally shared. It is also usually produced by public funding and in this sense already paid by the public. Free open data also fosters interoperability and creates a basis on which commercial applications can be built more easily. Trust and copyright issues are important, e.g., in Web 2.0 spirited social cultural portals, where end-users create, tag, and publish content of their own and the others'.

[11]http://linkeddata.org/
[12]http:/www.dbpedia.org/
[13]http://www.freebase.com/
[14]http://www.geonames.org/
[15]http://www.gutenberg.org/

1.5 OUTLINE OF THE BOOK

This book is an introduction to publishing CH contents on the Semantic Web as Linked Data. The idea is to provide a kind of cook book on how to create semantic portals of CH, where heterogeneous content is produced by a multitude of distributed organizations, and is harvested, harmonized, validated, and published as a service for human and machine users.

The text starts (Chapter 2) with presenting a motivating "business model" for this prototypical semantic portal scenario that can be considered a kind of standard model for publishing CH on the Semantic Web. In Chapter 3 requirements for publishing Linked Data are considered. The Semantic Web is based on the "layer cake model" of W3C that adds new standards above the XML[16] standard family, the *lingua franca* of the Web.

- *Metadata level.* The RDF data model[17] is the basis of the Semantic Web and Linked Data, and is used for representing metadata as well as other forms of content on the Web of Data. Metadata models for CH data are considered in Chapter 4.

- *Ontology level.* The RDF Schema and the Web Ontology Language OWL[18] are used for representing ontologies that describe vocabularies and concepts concerning the real world and our conception of it. Domain vocabularies and ontologies for CH are in focus in Chapter 5.

- *Logic level.* Logic rules, to be discussed in Chapter 6, can be used for deriving new facts and knowledge based on the metadata and ontologies. This can be used, e.g., to minimize cataloging work, make searching and browsing more effective, and to find serendipitous semantic links between CH objects.

After presenting technical foundations and models, issues related to annotating and harvesting CH content for a portal are presented in Chapter 7. Chapter 8 discusses intelligent services based on semantic linked data. The book is finally concluded in Chapter 9.

1.6 BIBLIOGRAPHICAL AND HISTORICAL NOTES

The idea of the World Wide Web (WWW) was proposed first in 1989 by Tim Berners-Lee, and more formally with Robert Cailliau in 1990. History of the early WWW is documented in the book "Weaving the Web" [14]. Already in the early days of the WWW the idea of a "Semantic Web," i.e., a web of machine interpretable data, has been around. However, the first generation of the WWW was targeted to humans, and was based on three simple technologies for mediating Web pages between human users: HTML, HTTP, and URLs.

From a scientific viewpoint, the Semantic Web is based on results of Artificial Intelligence, where semantic networks and logic-based knowledge representation have been studied from the

---

[16]http://www.w3.org/XML/
[17]http://www.w3.org/RDF/
[18]http://www.w3.org/2004/OWL/

late 50's; see, e.g., [126] for a thorough overview of this field. The first Semantic Web standard in use, Resource Description Framework (RDF), was published by W3C already in 1999, only a year after the XML recommendation. As another approach for the Semantic Web, Topic Maps [114] has been developed and published as the ISO standard ISO/IEC 13250:2003[19]. This standard is intended for the representation and interchange of knowledge, with an emphasis on the findability of information. The system originated from the idea of creating semantic indexes for publications. However, Semantic Web development really got off using the W3C standard stack after the publication of the seminal article "The Semantic Web" [15] in *Scientific American,* and the launch of the Semantic Web Activity at W3C.

The semantic technology did not penetrate the market as quickly as many other Web developments, say XML. A reason for this is complexity of some standards and their foundations in logic not so familiar in mainstream computing. In around 2005, the ideas on Linked Data and Web of Data started to gain momentum as a simple approach to the Semantic Web focusing on publishing large existing datasets, and using only simple RDF and lightweight ontologies. Combined with idea of Open Data, the idea of the Semantic Web has been adopted especially by the public sector [158], and several national initiatives have been started in the U.K.[20], U.S[21], and in smaller countries, such as Finland [67].

A thorough overview of Linked Data and Web of Data is presented in [53]. Semantic Web and linked data standards and technology, with pointers to related research and applications, can be accessed at W3C Web pages[22], and at the home pages of the Linked Data community[23]. The W3C Linked Library Data Incubator Group has evaluated the current state of library data management, outlined the potential benefits of publishing library data as Linked Data, and formulated next-step recommendations for library standards bodies, data and systems designers, librarians and archivists, and library leadership in a final report[24]. Another report "Linked Data for Libraries, Museums, and Archives: Survey and Workshop Report" with related goals was published at the same date, based on a workshop at the Stanford University[25]. Major international Semantic Web conferences include the International Semantic Web Conference (ISWC) and Extended Semantic Web Conference (ESWC). The World Wide Web conference (WWW) is the main yearly event for general Web research with a W3C focus.

A wide variety of Web applications in the museum domain have been presented in the proceedings of the Museums and the Web conference series since 1997, with papers available online[26]. The International Federation of Library Associations and Institutions (IFLA)[27] organizes a large annual World Library and Information Congress for libraries, and the International Council on

---

[19]http://www.isotopicmaps.org/
[20]http://data.gov.uk/
[21]http://www.data.gov/
[22]http://www.w3.org/standards/semanticweb/
[23]http://linkeddata.org/
[24]http://www.w3.org/2005/Incubator/lld/
[25]http://www.clir.org/pubs/abstract/reports/pub152
[26]http://www.archimuse.com/conferences/mw.html
[27]http://www.ifla/

Archives (ICA)[28] has a similar annual congress series, International Conference of the Round Table on Archives (CITRA) for archivists.

The intersection of computing and the disciplines of the humanities are studied in the field of *Digital Humanities*, also called *Humanities Computing*. [105] The general goal here is to develop and apply computational methods in humanities research. Since 1990, the digital humanities community has been organizing the Digital Humanities conference series[29]. A major journal in the field is the *Digital Humanities Quarterly*[30].

---

[28]http://www.ica.org/
[29]http://digitalhumanities.org/conference
[30]http://www.digitalhumanities.org/dhq/

CHAPTER 2

Portal Model for Collaborative CH Publishing

This section presents a prototypical "business model" and an architecture for publishing CH collections and other materials collaboratively on the Semantic Web as linked data. Benefits from the end-users' and publishers perspectives are analyzed, and challenges discussed.

2.1 GLOBAL ACCESS FOR LOCAL LINKED CONTENT

Contents of memory organizations are usually situated in different data "silos": the databases are distributed at different locations and use different database systems and schemas. Each organization typically publishes its content using custom made home pages and collection browsers. Organizational boundaries of data create severe obstacles for information retrieval to end-users who have to know what content could be found in what collection, have to learn how to use different search systems, and have to query and collect data multiple times using different Web interfaces. The end-user rather needs a single global service for accessing data situated in different museum databases than a set of local organization specific systems. To facilitate this, local contents need to be aggregated and combined somehow, and then used via a harmonized user interface, a Web portal.

Combining data from different sources and providing them through a unified view is studied in the field of *data integration*. Data integration can take place at different phases of information retrieval. There are two main strategies available: 1) integrate data dynamically while processing queries or 2) integrate data beforehand in a separate pre-processing stage.

2.1.1 FEDERATED SEARCH

In the former case, a search query can be sent to distributed local databases and the results be combined into a global hit list. This approach is called *federated search*, also known as *multi-search* or *meta-search*. Federated search consists of the following steps.

1. Transform the query for each participating database service and broadcast it.

2. Collect the results from the services and merge them into one result set.

3. Present the result set in a concise format with minimal duplication.

4. Provide the end-user with means for inspecting the merged result set, e.g., by sorting it in different ways.

A major benefit of federated search is that it requires only loose coupling of the collection database systems involved: the participating organizations must only agree upon the querying protocol, and the database systems can continue their independent lives as before. Due to its simplicity, federated search has been applied widely in CH (and other) portals, and there are many national cultural collection systems on the Web based on it, such as the portals Collections Australia Network[1] [140], Artefacts Canada[2], and the Swedish Samsök[3]. The approach is especially useful when searching for objects of similar kind represented using similar metadata formats, such as books in libraries.

Federated search can be performed using two strategies. In the *Local as View* (LAV) strategy local databases provide the search system with their own local views, onto which the federated query must be transformed. In the *Global as View* (GAV) strategy, a shared global view is mediated to local services to be answered. Software modules for doing the needed querying transformations are called *wrappers*.

A fundamental limitation of federated search is that by processing the query independently at each local database, the global dependencies, i.e., associations between objects across *different* collections are difficult to found. Since finding semantic links between collection items is one of the main goals of linked data, a semantic CH portal for linked data cannot be based on the federated search paradigm without sacrificing much of the potential of the Semantic Web [63].

2.1.2   DATA WAREHOUSING

The alternative approach to federated search is to consolidate first the local collections into a global repository, a *data warehouse*, and then search the global database. Mutually shared conceptual models, ontologies, can be used for enriching the contents and for making the collections interoperable. To show the associations to the end-user, the collection items can be represented as Web pages interlinked with each other through the semantic associations. A major challenge in this approach is that a separate content creation process is needed for consolidating the global repository based on local databases, and maintaining it as the participating databases evolve. More collaboration and coordination is needed between the content providers of such a portal.

2.2   COLLABORATIVE PUBLISHING OF LINKED DATA

Linked Data of CH is heterogeneous and at the same time semantically interlinked, as illustrated in Figure 2.1. For example, the content may contain a person's narrative biography, works of art she created, places of interest where she lived in, Wikipedia articles or novels about her or by her, social connections to relatives and other persons, and historical events that the person was involved with. Such multi-perspective content is likely to be created by different organizations and individual citizens independently from each other, using different metadata schemas, vocabularies,

---

[1]http://www.collectionsaustralia.net/
[2]http://www.pro.rcip-chin.gc.ca/artefact/index-eng.jsp
[3]http://www.kb.se/libris/samsok/

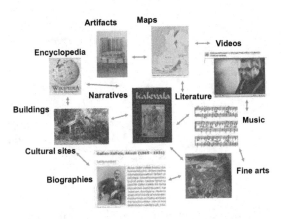

Figure 2.1: Cultural heritage is semantically heterogeneous and mutually linked.

cataloging conventions, and languages, as illustrated in Figure 2.2. Linking heterogeneous data in such a distributed content provision environment is challenging to end-users of cultural data, and to organizations and communities producing the contents.

Figure 2.2: Cultural heritage content is produced by independent actors, typically without much coordination.

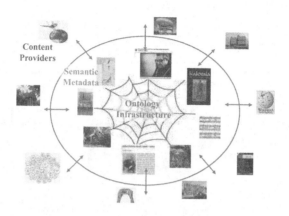

Figure 2.3: A model for Linked CH publishing, based on a shared ontology infrastructure in the middle.

Fortunately, the ideas of the Semantic Web and Linked Data can be applied to address the problems of data interoperability and distributed content creation at the same time, as illustrated in Figure 2.3. Here the publication system is illustrated by a circle. A shared semantic network, an ontology infrastructure, is situated in the middle and consists of shared aligned ontologies and metadata models used in describing the contents of the system. If content providers outside of the circle provide the system with metadata about CH using the same concepts, the data is automatically linked with each other and forms a GGG.

For example, if metadata about a painting created by Picasso comes from an art museum, it can be enriched (linked) with, e.g., biographies from Wikipedia and other sources, photos taken of him, information about his wives, books in a library describing his works of art, related exhibitions open in museums, and so on. The metadata about the painting is semantically enriched, if linking can be established correctly. At the same time, the contents of any organization in the portal having Picasso-related material get enriched by the metadata of the new artwork entered in the system. This is clearly a win-win situation for everybody to join such a system; collaboration pays off.

The potential benefits are especially high in situations where CH content related to a topic is distributed in different places, and getting a holistic view of it becomes more challenging. For example, in Europe there are lots of different countries whose border lines have changed during history, and the CH content is distributed to different national collections. In the same way, much of the CH content of former colonies, say treasures of ancient Egypt, have been transported to Western museums, and are no longer available at the original locations. Using a system such as the one in Figure 2.3 CH collections could be united at least virtually on the Semantic Web.

2.3 BENEFITS FOR END-USERS

A semantic portal application, following the model illustrated above, is useful from the end-users' viewpoint in several ways (to be discussed in more detail in Chapter 8.

- *Global view to heterogeneous, distributed contents.* The contents of different content providers can be accessed through one service as a single, seamless, and homogeneous repository. Only a single user interface has to be learned.

- *Automatic content aggregation.* Satisfying an end-user's information need often requires *aggregation* of content from several information providers. For example, when looking for data about an artist, relevant information may be provided by museum collections, libraries, archives, authority records, ontologies, and other sources.

- *Semantic search.* In traditional portals, search is usually based on free text search (e.g., Google), database queries, and/or a stable classification hierarchy. Semantic content makes it possible to provide the end-user with more "intelligent" facilities based on ontological concepts and structures, such as *semantic search* , *semantic autocompletion*, and *faceted search*.

- *Semantic browsing and recommendations.* Semantic content also facilitates semantic browsing and recommendations for additional information. Here semantic associations between search objects can be exposed to the end-user as recommendation links, possibly with explicit explanations.

- *Other intelligent services.* Also other kinds of intelligent services can be created based on machine interpretable content, such as knowledge and association discovery, personalization, and semantic visualizations based on, e.g., historical and contemporary maps and timelines.

2.4 BENEFITS FOR PUBLISHERS

Semantic collaborative portals are also very attractive from the content publishers viewpoint.

- *Distributed content creation.* Portal content is usually created in a centralized fashion by using a content management system (CMS). This approach is costly and not feasible if content is created in a distributed fashion by independent publishers, e.g., by different museums and other memory organizations. Semantic technologies can be used for harvesting and aggregating distributed heterogeneous content into global content portals.

- *Automated link maintenance.* The problems of maintaining links up-to-date is costly from a portal maintenance viewpoint. In semantic portals, links can be created and maintained automatically based on the metadata and ontologies.

- *Shared content publication channel.* In the cultural domain the publishers usually share the common goal of promoting cultural knowledge in public and among professionals. A semantic portal can provide the participating organizations with a shared, cost-effective publication channel.

• *Enriching each other's contents semantically.* Interlinking content between collaborating organizations enriches the contents of everybody "for free."

• *Reusing aggregated content.* The content aggregated into a semantic portal can be reused in different applications and cross-portal systems.

The model presented is compatible with the idea of end-user created CH content on the Web 2.0, successfully applied in portals such as Steve Museum[4] and Powerhouse Museum[5].

2.5   NEW CHALLENGES

Production and utilization of semantic data also poses new challenges. Obtaining interoperability requires in practice more disciplined use of standards, harmonized metadata models, shared vocabularies, and shared best practices. The main challenge is often rather organizational than technical: changing, e.g., cataloging practices is not easy, and if changes are made, there is the question of what to do with already cataloged legacy metadata. A practical difficulty is that content management systems in use do not support creation of Semantic Web data. Furthermore, when aggregating content across organization boundaries more collaboration and harmonization of data is needed.

Developing more intelligent applications also sets new challenges: such systems are typically more complex, require specific skills from programmers, new tools, and are typically also computationally more complex and not necessarily scale up so easily. Data enrichment via linked data is promising, but in practice the datasets available have quality problems: many of them, such as DBpedia, have been produced automatically by machines without human touch. A problem here is that the URI identifiers used for concepts (e.g., persons and places) in different datasets are typically different, and the data mappings are not complete or contain errors. Vocabularies used may contain loops, missing labels, violate semantic constraints of standards, and so on. Finally, when reusing content as services one becomes dependent of an external system controlled by somebody else, and often have to be content with suboptimal API services, quality of service, licensing policies, etc.

However, it is clear that these challenges need to be tackled in one way or another when integrating collection data on a semantic level. Semantic Web technologies provide a standard approach and a tool set that has already been successfully applied for the task, so why not try it instead of starting from scratch and possibly ending up reinventing similar solutions again?

2.6   COMPONENTS OF A SEMANTIC PORTAL SYSTEM

A collaborative semantic portal for CH needs three major components depicted in Figure 2.4.

---

[4]http://www.steve.museum/
[5]http://www.powerhousemuseum.com/collection/database/

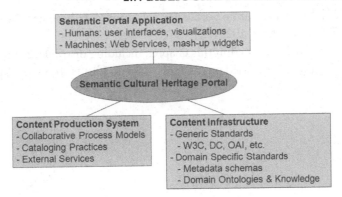

Figure 2.4: Components of a semantic CH publishing portal.

1. **Content Infrastructure.** The domain ontologies include a set of cross-domain ontologies for general concepts (e.g., artifact types and materials), authorities (persons, groups, and organizations), places (current and historical), time periods, and events. Without sharing references and identifiers (URIs in Web context) of mutually agreed indexing concepts, or mappings between indexing schemes, data cannot be linked with each other.

2. **Content Production System.** In addition, content creation models and best practices are needed for guiding and harmonizing cataloging. Furthermore, models and protocols for harvesting content from organizations and individual citizens are needed, and methods for maintaining harvested linked data need to be developed. For example, if an error in data is found, it should be corrected not only in the portal but also in the original collection database.

3. **Semantic Portal System.** The portal itself is the most visible part of the system, but only the tip of an iceberg. It provides the human end-user with semantic searching and browsing services and visualizations of the content. For the machine, i.e., other applications on the Web, such as the Web portals of the participating organizations, the system is accessible via APIs.

In the following chapters, these three components are discussed in more detail. First, general requirements for publishing linked data are presented. After this, issues related to the portal content infrastructure, including metadata models, ontologies, and knowledge representation based on rules are discussed. After this, the content production process is in focus, and finally services of the portal to human and machine users.

2.7 BIBLIOGRAPHICAL AND HISTORICAL NOTES

Portal types include *service portals* collecting a large set of services together (e.g., Yahoo! and other "start pages"), *community portals* [138] acting as virtual meeting places of communities, and *information portals* [120] acting as hubs of data. Early discussions of semantic information portals include [94, 120]. Several semantic portals for collection data have been developed and many are still online [6, 7, 13, 63, 72, 99, 132, 152].

The portal model presented in this chapter, including distributed content creation based on an ontology infrastructure, in based on the cross-domain CultureSampo system[6] [72, 99].

---

[6]http://www.kulttuurisampo.fi/

CHAPTER 3

Requirements for Publishing Linked Data

This chapter first presents general technical requirements for publishing CH content as Linked (Open) Data, based on five quality levels. After this, standard APIs for Linked Data repositories are presented. Based on such APIs, end-user applications such as semantic portals can be built in a disciplined manner. Finally, quality issues of linked data are discussed.

3.1 FIVE-STAR MODEL FOR LINKED DATA

Data can be published on the Web at increasing levels of openness and linkage, as characterized by the so called five-star system[1] presented in Table 3.1.

Table 3.1: Five Stars System for Evaluating Linked Data	
★	**Data Structure**. Data is available as structured data.
★★	**Licensing**. Data is available on the Web (in whatever format) under an open license.
★★★	**Format**. Non-proprietary open formats are used, e.g., CSV (Comma Separated Values) format instead of Excel's own proprietary format.
★★★★	**Identifiers**. HTTP URIs are used to identify things, so that people can point to the data and serve RDF from it.
★★★★★	**Data Linking**. Data is linked internally and externally to other data to provide context.

The general goal is to try to earn as many stars as possible, one star for each point listed in the table. Below, guidelines for creating linked data publications are presented star by star, starting from the requirement for structured data.

3.1.1 PUBLISHING STRUCTURED DATA

Publishing data in structured formats means that the data cannot only be read by humans but also interpreted and reused by computers for applications. For example, publishing an Excel sheet in CSV form is much more reusable than publishing an image scan of a table, even if the image may be of use for human readers.

[1]http://www.w3.org/DesignIssues/LinkedData.html

The first star is earned by using any structured formats, but for Linked Data the optimal choice is to use the Resource Description Framework RDF[2]. RDF has many syntaxes, and the choice depends on the use-cases in mind. Textual syntaxes of RDF are called *serializations* because text represents multi-dimensional graphs as a series of characters in a file.

1. **RDF/XML**[3] is the original standard way of representing RDF graphs. It is based on XML and can be parsed and validated easily by most RDF tools. For example, the metadata about the Pablo Picasso resource *p-4* in Figure 1.3 can be represented in RDF-XML form as:

```
<?xml version="1.0?''>
<rdf:RDF xmlns:rdf="http://www.w3.org/1999/02/22-rdf-syntax-ns#"
        xmlns:ex="http://example.org/">
   <rdf:Description rdf:about="http://example.org/p-4">
     <rdf:type rdf:resource="http://example.org/Person"/>
     <ex:name>Pablo Picasso</ex:name>
     <ex:birth-place rdf:resource="http://example.org/p-18"/>
     <ex:birth-time>1881</ex:birth-time>     </rdf:Description>
</rdf:RDF>
```

RDF/XML syntax is verbose to write and difficult to read for human users. Therefore, simpler syntaxes N-triples, N3, and Turtle have been developed.

2. **N-Triples** is a straightforward way of serializing RDF graphs as a set triples. It is a simplified version of *Notation 3 (N3)*[4], a human-friendly notation and extension for RDF graphs. N-Triple format lists one full triple $< subject, predicate, object >$ in one line. It is useful when creating datasets that can be loaded efficiently into a triplestore triple by triple without loading the whole data into memory. For example, the metadata about the resource *p-4* in Figure 1.3 in N-triples format is presented below:

```
@prefix ex: <http://example.org/> .
@prefix rdf: <http://www.w3.org/1999/02/22-rdf-syntax-ns#> .

ex:p-4 rdf:type ex:Person . ex:p-4 ex:name "Pablo Picasso" .
ex:p-4 ex:birth-place ex:p-18 . ex:p-4 ex:birth-time 1881 .
```

3. **Terse RDF Triple Language Turtle**[5] is arguably the simplest format to use from a human perspective, and has become very popular in Linked Data. It is compatible with N-Triples, but is terser to write. It is also compatible with the query pattern syntax used in SPARQL[6], the query language for the Semantic Web.

For example, the metadata about the resource *p-4* in Figure 1.3 in Turtle can be represented without repetition of the subject in this way:

---

[2]http://www.w3.org/RDF/
[3]http://www.w3.org/TR/REC-rdf-syntax/
[4]http://www.w3.org/TeamSubmission/n3/
[5]http://www.w3.org/TR/turtle/
[6]http://www.w3.org/standards/techs/sparql#w3c_all

```
@prefix rdf: <http://www.w3.org/1999/02/22-rdf-syntax-ns#> .
@prefix ex: <http://example.org/> .

ex:p-4 rdf:type ex:Person ;   ex:name "Pablo Picasso" ;
  ex:birth-place ex:p-18 ;   ex:birth-time 1881 .
```

4. **RDFa**[7] is a mark-up scheme for embedding RDF descriptions in XHTML and HTML5[8]. RDFa provides a mechanism for publishing metadata on Web pages, from which they can be extracted and harvested by, e.g., search engines. For example, in the following HTML paragraph, the resource *p-4* is given two RDF properties using the literal text of the document:

```
... <p about="http://example.org/p-4">
  In his paintings <span property="ex:name">Pablo Picasso</span>,
  born in  <span property="ex:birth-time">1881</span>,  expressed ...
</p> ...
```

The GRDDL mechanism for Gleaning Resource Descriptions from Dialects of Languages[9] can be used for extracting RDF out of a Web page.

RDFa is the Semantic Web version of the idea of 1) *microformats*[10] and 2) *microdata*. In all three approaches, developed by different communities, the underlying idea is the same: to embed structured data for machines to use in human readable HTML pages on the Web. By embedding, the amount of markup one needs to write can be reduced, because one does not need to write fully separate HTML and structured metadata descriptions.

The main differences between the approaches are: Microformats are based on using traditional HTML tags, not specifically intended for the purpose of representing metadata, in such ways that the Web pages can still be rendered by browsers without problems for human usage. Microdata, on the other hand, uses specific tags of HTML5 for representing structured data, and provides a collection of shared vocabularies that webmasters can use to mark up their pages in ways that can be understood by search engines.

A most important specification for embedding metadata in XHTML for search engines is Schema.org[11], supported by major search engines, such as Google, Bing, Yahoo, and Yandex. The specification is Semantic Web spirited including ontology-like shared vocabularies for annotating contents. A major difference between Schema.org, developed by industries, and RDFa, developed by W3C, is that RDFa is based on using URIs, making it more compatible with Linked Data. However, it is possible to map Schema.org descriptions into corresponding representations in RDF. Schema.org vocabularies are available in different RDF formats, JSON[12], and CSV, too[13]. JSON

---

[7]http://www.w3.org/TR/xhtml-rdfa-primer/
[8]http://dev.w3.org/html5/spec/single-page.html
[9]http://www.w3.org/TR/grddl-primer/
[10]http://microformats.org/
[11]http://schema.org/
[12]http://www.json.org/
[13]http://schema.rdfs.org/

(JavaScript Object Notation) is an easy to use lightweight data-interchange format that has become very popular on the Web and in Linked Data[14].

3.1.2 OPEN LICENSING

In linked data publishing the notion of openness is advocated along two major dimensions. Firstly, by promoting open licensing policies and secondly, using open formats instead of closed proprietary formats. Data linking technologies can be applied, however, to both open and closed datasets.

Licensing and open data are a tricky issue in many areas of the CH domain, where copyright of creative works and privacy issues can be quite thorny, and the interests of content creators and publishers can be conflicting. For example, the copyrights of a movie may be shared by script and film makers, producers, actors, and other parties involved, and at the same time distribution rights for TV, movie theaters, and other channels may be owned by different organizations. A big question is who is going to finance authors of creative works if contents were open and free, and there were no royalties of sales available. In addition, there may be privacy issues concerning the people presented on artworks. With many historical materials it may be practically impossible to find out who, if anybody, could possibly have rights to a piece of content, say a photograph, or whose privacy could be at risk, if the photograph is published.

The general motivation for opening data for free use is that although it may be beneficial for an individual organization to charge for its content, the situation may be vice versa when considering the situation from a larger, say a national perspective. In CH the general underlying goal is to maintain tradition and enrich and distribute knowledge about it, which makes the domain particularly suitable for opening data.

Along the political Open Data movements in different countries, the current trend is clearly toward open publishing policies. For example, the Open Government License for Public Sector Information[15] that is used in the HM Government linked data initiative in the U.K. contains the following rules.

1. You are free to copy, publish, distribute and transmit the Information.

2. You can adapt the Information.

3. You can exploit the Information commercially for example, by combining it with other Information, or by including it in your own product or application.

However, one should "acknowledge the source of the Information by including any attribution statement specified by the Information Provider(s) and, where possible, provide a link to this license." If this for some reason is not feasible, some additional options are suggested. Information is provided on an "as is" basis without warranty, and there are exemptions mentioned, e.g., that personal data, patents, and trademarks are not under the license.

---

[14]http://www.w3.org/community/json-ld/
[15]http://www.nationalarchives.gov.uk/doc/open-government-licence/

A more and more used licensing model on the Web is the Creative Commons model[16] (CC) used, e.g., in the Linked Data Publication of Europeana[17]. CC licenses are created easily and interactively using a Web tool in multiple languages by making few choices to questions, such as: Do you allow modification of the work? Is it possible to use the work for commercial purposes? An often used model in Linked Open Data is the Attribution License requiring essentially only the acknowledgment of the original creators of the content. This means that the content is free for anyone to share, copy, modify, distribute, transmit, and to remix the data to adapt the work, including commercial uses.

Linked Open Data is usually available free of charge on the Web. However, "open" is not the same thing as "free (of charge):" data can be opened and still have a price tag on it.

3.1.3   OPEN FORMATS

Using open formats[18], such as XML, Rich Text Format (RTF), or Open Document Format for Office Application (ODF), instead of proprietary formats is useful for many reasons.

- Open standards foster interoperability between tool makers, application developers, and content providers. Everybody is able to access and read open formats.

- Using open standards makes it possible to avoid vendor lock and to, e.g., change software or service provider in case of bankruptcy.

- Open standards are more secure for long time preservation of digital materials, which is particularly important for CH institutions.

- Open standards are free of royalty charges.

- Open standards are good for virus protection.

3.1.4   REQUIREMENTS FOR IDENTIFIERS

A key concept on the Semantic Web is the notion of *resource*. A resource is any identity we want to represent metadata about. Identities, i.e., resources, are referenced to by Universal Resource Identifiers (URI).

The idea is that resources are identified by HTTP-based Web identifiers for machine use, not by their names or titles in a particular natural language that are used for human consumption in applications. Using URIs makes it possible to differentiate entities with similar names and decouple machine semantics from human languages. This enables, e.g., language-independent multilingual applications.

---

[16] http://creativecommons.org/
[17] http://pro.europeana.eu/linked-open-data
[18] http://www.openformats.org/

HTTP URIs and IRIs

There are two kinds of URIs that should not be confused with each other:

1. URIs identifying *real-world* or imaginary *things*, say "London," "chair," "World War I," or "unicorn," and

2. URIs referring to *Web documents*, say the homepage of London, a Web page about a museum collection item, or its representation in RDF form.

For example, let us consider minting a URI for the World War I (WWI) event "Battle of Albert" in France in 1914. When indexing data about it, it is not clear what language should be used. For example, "Bataille d'Albert 1914" in French or "Albertin taistelu 1914" in Finnish could be used in addition to the English title. The solution is to use neutral URIs for the underlying concept, such as

```
http://dbpedia.org/resource/r3981672
http://dbpedia.org/resource/Battle_of_Albert_(1914)
```

This latter URI is actually used in DBpedia for this particular battle, and although it is based on English, it is language neutral in the sense the URI indeed has different labels "Bataille d'Albert (1914)" in French and "Battle of Albert (1914)" in English attached to it, and separated from the URI. The former URI would be more neutral in terms a languages, but on the other hand also more difficult to use from a human perspective.

URIs are constructed using a limited subset of the ASCII character set. URL encoding, also known as *percent encoding*, can be employed for including *reserved characters* in URIs. This means that the reserved character is replaced by the % character followed by two hexadecimal digits. For example, the question mark '?' has a special meaning in the URI syntax as the separator of the query part, but can be used in the URI string without this meaning by encoding it as %3F.

The notion of URI has been generalized into the notion of *Internationalized Resource Identifier* (IRI). IRIs may contain characters from the Universal Character Set (Unicode/ISO 10646), including Chinese or Japanese kanji, Korean characters, Cyrillic characters, and so on. HTTP IRIs are also sometimes used as identifiers for concepts. Typing in IRIs sets special requirements for the input device, such as a keyboard.

Content Negotiation and Redirection

A fundamental principle of Linked Data is that HTTP URI identifiers not only identify things uniquely on a WWW scale, but are also addresses for getting more information about the resources. This idea is similar to URLs that are used for identifying and retrieving Web pages.

Information about a URI for a real world thing is looked up by either 1) Semantic Web client applications or by 2) Web browsers for human consumption. In order to support both use cases, Web servers use the *content negotiation* mechanism of the HTTP protocol[19]. An HTTP request sent by a

---

[19]http://www.ietf.org/rfc/rfc2616.txt

browser or a software agent includes a textual header that indicates what data formats and languages it prefers. For example, the HTTP header below indicates that a client requester wants an HTML or XHTML representation of a chair in a collection, identified by the URI `http://www.museum.org/collection/chair-34`, in English or German:

```
GET /collection/chair-34 HTTP/1.1 Host: www.museum.org
Accept: text/html, application/xhtml+xml Accept-Language: en, de
```

The server then selects the best match from its file system or generates the desired content on demand, and sends it back to the client. The answer for the GET request above could be, for example

```
HTTP/1.1 200 OK Content-Type: text/html
Content-Language: en
Content-Location: http://www.museum.org/collection/chair-34.en.html ...
```

followed by the English HTML page about the chair. Using content negotiation, the client can be served with either an HTML page illustrating a real world concept or its RDF representation, depending of the request header used.

In this example, content negotiation returned immediately the desired content. It is also possible to *redirect* the request to another address, by using a special 302 or 303 status code in the response:

```
HTTP/1.1 303 See Other
Location: http://www.other-museum.org/collection/chair-75.en.html
```

This makes it possible for a server to direct the client's URI request to another address or an RDF representation of it, depending on the case.

Content negotiation and redirection can be used for making a difference between the concept reference, RDF information about the concept, and an HTML Web page about it. For example, in DBpedia there are actually three different URIs related to the Battle of Albert (1914) resource:

1. **the concept** URI:

   `http://dbpedia.org/resource/Battle_of_Albert_(1914)};`

2. **the RDF representation** of the URI:

   `http://dbpedia.org/data/Battle_of_Albert_(1914);`

3. **the HTML page** showing the concept RDF data:

   `http://dbpedia.org/page/Battle_of_Albert_(1914).`

Depending on the HTTP request header, the DBpedia URI of the Battle of Albert can be resolved into an RDF document or an HTML page rendering it, and the corresponding Web document is returned. The resource URI is redirected into the HTML page if it is written into a browser.

Providing needed content negotiation and redirecting services for linked data URIs is a responsibility of the data publisher of the URI domain name, here dbpedia.org. There are services available, such as a Persistent Uniform Resource Locator service maintained by OCLC[20], by which permanent identifiers can be minted and redirected to different physical servers in coherence with the dynamic and changing Web infrastructure. The act of retrieving a representation of a resource identified by a URI is known as *dereferencing*.

Using URIs has many obvious benefits for linking data. If different content providers index their data using shared URIs, then the distributed data can be linked together automatically in order to enrich it. For example, if the DBpedia concept URI above is used for indexing documents and objects related to the Battle of Albert 1914 in museums or libraries, then these collection items can be linked together, to Wikipedia articles in different languages about the Battle of Albert, and to other resources related to the URI via, e.g., the Linked Open Data cloud. The global domain name infrastructure of the WWW guarantees that only the owner of the domain (here dbpedia.org) can introduce synonymous URIs, which prevents semantic confusion on a global scale.

One challenge in using URIs is that there are often several URIs for a concept already in use. Using a single global ontology for URIs would be an optimal solution, but the reality is that different repositories and communities will continue using different identifiers for the same things. For example, there may be data about WWI or the Battle of Albert in Freebase or in the Imperial War Museum databases, national land surveying organizations have their own established identifiers for places, and so on. The LD solution to address this issue is to create ontology alignments (mappings) between repositories. They define what URIs refer to the same concept in different RDF stores, or overlap in meaning when the concepts do not fully correspond to each other.

Minting URIs

Another challenge is that there are different ways of constructing URIs, so how does one determine what naming policy to use? The two main technical choices to use are[21]:

1. **Hash URIs**, such as `http://example.org/about#p-4`, show the '#' character before the local name. In the HTTP protocol, the hash and the local name is removed, and a single request for all resources in the document is made, in this case `http://example.org/about`.

2. **303 URIs**, such as `http://example.org/about/p-4`, can be configured to return individual RDF descriptions, using redirection.

---

[20]`http://purl.org`
[21]`http://www.w3.org/TR/cooluris/`

Using 303 URIs and redirection may be preferable when retrieving RDF of resources in very large RDF files. This avoids downloading large RDF documents, but at the cost of making an additional HTTP request due to redirection.

A URI typically contains a path starting from the domain name and ending up in a local name. There are different strategies for creating such names as well as the corresponding URLs to be resolved. For example, the DBpedia example above illustrated one possibility. In W3C's own recommendation documents the publication date is used in the path in order to separate different publications of a recommendation. The strategy depends on the content and case at hand.

A fundamental principle for minting URIs is that "cool URIs do not change"[22]. Therefore, it is a good policy to use URIs that are neutral in terms of meaning and language. This can be a problem, e.g., in the DBpedia URI for the Battle of Albert 1914, since the URI is based on English, and if it is later discovered that the Battle of Albert actually took place in a year other than 1914. On the other hand, simple mnemonic names with a meaning make URIs easier to use from a human perspective. Again, there are no single solutions and one has to follow one's nose remembering that any later change in a URI needs to be taken care of in systems already using the old URI.

Various identifier schemes are in use in the CH domain. For example, libraries use International Standard Book Number (ISBN) codes for books, such as 978-3-540-70999-2. There is the International Standard Serial Number (ISSN) code for periodical publications, such as magazines, and the scope of International Standard Name Identifiers (ISNI)[23] is the identification of public identities of parties. Some organizations and systems use Uniform Resource Name (URN) identifiers. There are actually tens of specific URI schemes developed for representing identifiers, such as ISBN and URN, registered by the Internet Assigned Numbers Authority (IANA)[24]. However, although these identifiers are URIs, they are not HTTP URIs based on the HTTP URI scheme. This means that such identifiers do not carry along the information about the Web location in which additional information about them can be obtained. As a result, they fail to earn the fourth star in the LD requirements (Table 3.1), and are not recommended for Linked Data as a resource identification mechanism. It is possible, of course, to use identifiers such as ISBN as part of HTTP URIs, e.g., as local names.

3.1.5 LINKING DATA INTERNALLY AND EXTERNALLY

In an RDF triple $< S, P, O >$, the subject S and predicate P must be resources identified by URIs[25], while the object O can be either a resource or a literal data value (string, number, date etc.). A literal value, e.g., the name or age of a person, cannot have further properties of its own. Only if the object O is an URI, then a data link between the two resources S and O is established.

---

[22]http://www.w3.org/Provider/Style/URI
[23]ISO 27729 standard, http://www.isni.org/
[24]http://www.iana.org/assignments/uri-schemes.html
[25]Some resources in RDF may be anonymous *blank nodes* (*bnodes*). They are unique resources whose identifiers are managed internally by the RDF system.

A link can connect two resources within a dataset, or two resources from separate datasets. The power of the Linked Data approach is based on creating rich data links of both kinds. It is easy to create RDF datasets of resources with only literal properties, but such datasets are not really Linked Data but only a syntactic RDF variant of the original data. The proportion of data link triples in a data set as well as the proportion of internal and external links can be used as measures of linkedness of a dataset.

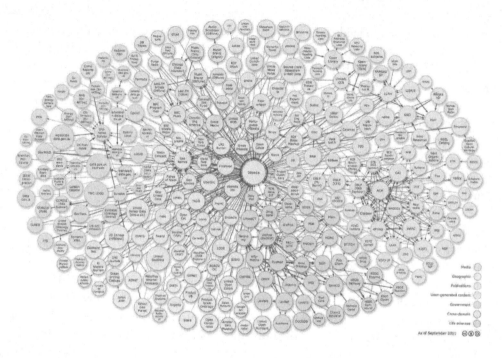

Figure 3.1: The Linked Open Data Cloud (here as of September 2011) consists of datasets (depicted as bubbles) in different domains (identified by coloring), and mappings between similar resources in the datasets (illustrated by arcs between the bubbles).

Many datasets use independent URI minting policies of their own, and a concept often has several copies of itself in different datasets with different URIs. For example, same places and persons often appear in different collections with different identifiers. In order to link such resources properly, sets of external links are usually created, mapping datasets onto each other. For example, Figure 3.1 depicts the Linked Open Data cloud[26] where each bubble is a dataset. An arch between two datasets A and B denotes a mapping M between them and is represented as a set of data links (a set of triples). Typically, `owl:sameAs` triples $< R_A, owl : sameAs, R_B >$ are used in mappings, where R_A is a resource in A and R_B in B. Many triplestore systems can automatically treat resources

[26]Diagram courtesy of Richard Cyganiak and Anja Jentzsch, http://lod-cloud.net/

R_A and R_B as if they were the same, hiding mapping details from the user. Also other kinds of predicates can be used in dataset mappings, such as `rdfs:subClassOf` or `rdfs:subPropertyOf`, establishing hierarchies of classes and properties, respectively. In the SKOS vocabulary standard[27], inexact mappings between vocabulary terms can be established using mapping properties, such as `skos:closeMatch`. The datasets can be merged with each other by simply loading A, M, and B into a single RDF graph.

3.2 REQUIREMENTS FOR INTERFACES AND APIS

A linked RDF dataset is published by making it available via interfaces. Typically, the following interfaces are provided.

1. **Linked Data Browsing**. Linked Data browser interface based on URI dereferencing. This interface makes the data set browsable by the various tabulators and LD browsers created for the Web of Data.

2. **SPARQL endpoint**. Published for querying the data in a standard way for, e.g., mash-up applications.

3. **Download facility**. Possibility of downloading the data as an RDF data dump is provided.

4. **Human interfaces**. Human end-user interfaces for searching and browsing the data.

3.2.1 LINKED DATA BROWSING

In Linked Data browsing each resource is rendered by showing its literal and resource values as properties. Resources are seen as HTML links; a resource is rendered by clicking after which browsing can go on. Several LD browsers have been created such as the Tabulator[28], Sig.ma[29], OpenLink Data Explorer[30], Disco[31], Ontology-browser[32], and Falcons Explorer[33] to name a few.

3.2.2 SPARQL ENDPOINT

SPARQL Protocol and RDF Query Language (SPARQL)[34] is the standardized query language for RDF data. It bears some similarities with SQL for relational databases. There are four query forms available.

1. **SELECT query** is based on a graph pattern, represented using Turtle and query variables beginning with '?' for resources. The pattern is matched with the underlying RDF graph,

---

[27]http://www.w3.org/2004/02/skos/
[28]http://www.w3.org/2005/ajar/tab
[29]http://sig.ma/
[30]http://linkeddata.uriburner.com/ode/
[31]http://linkeddata.uriburner.com/ode/
[32]http://code.google.com/p/ontology-browser/
[33]http://ws.nju.edu.cn/explorer/
[34]http://www.w3.org/TR/rdf-sparql-query/

resulting in different variable bindings for the variables that match the pattern. The bindings are the results for the query and are returned in table form. For example, the following SPARQL query would return a table of two columns listing the names and corresponding materials of all instances of the class Artifact.

```
PREFIX ex: <http://museum.org/artifact/>

SELECT ?name ?material WHERE {
    ?artifact a ex:Artifact .   ?artifact ex:name ?name .
    ?artifact ex:material ?material . }
```

2. **CONSTRUCT query** is used to extract information from the SPARQL endpoint, and to transform the results into valid RDF.

3. **ASK query** only checks whether a given graph pattern matches the underlying RDF graph, and returns *true* or *false* as result.

4. **DESCRIBE query** is used to read the RDF description of a given URI.

A SPARQL endpoint provides external users with a dynamic service for information retrieval over the HTTP protocol without the need to copy and transfer the dataset. A variety of SPARQL endpoint software packages is available, including both open source and commercial systems, such as 4Store[35], AllegroGraph[36], Jena[37], Sesame[38], and Virtuoso[39].

A list of currently alive SPARQL endpoints is maintained on W3C wiki pages[40].

3.2.3   DOWNLOAD FACILITY

The download facility is useful for, e.g., external users willing to analyze, modify, or enrich the dataset, or use it in offline and other applications.

3.2.4   HUMAN INTERFACES

Finally, various application interfaces for the dataset can be created for human users, such as ontology browsers in ontology library systems [112]. Several search engines for searching RDF data has been created, such as Falcons[41], Sindice[42], Swoogle[43] [39], SWSE[44] [59], and Watson[45] [31].

[35] http://4store.org/
[36] http://www.franz.com/agraph/allegrograph/
[37] http://jena.apache.org//
[38] http://www.openrdf.org/
[39] http://virtuoso.openlinksw.com/
[40] http://www.w3.org/wiki/SparqlEndpoints
[41] http://ws.nju.edu.cn/falcons/objectsearch/index.jsp
[42] http://sindice.com/
[43] http://swoogle.umbc.edu/
[44] http://swse.org/
[45] http://watson.kmi.open.ac.uk/WatsonWUI/

3.3   BIBLIOGRAPHICAL AND HISTORICAL NOTES

The five-star model was proposed and discussed by Tim Berners-Lee[46]. The RDF data model and RDF/XML syntax was published by W3C first in 1999 and revised in 2004. The use of the human readable concise Turtle notation was developed from N3 and N-Triple notations. It is now used widely and is being standardized. Information about RDF-based standards and tools can be found at W3C pages[47].

Microformats are being developed and advocated by the microformats.org community[48], and the use of open formats by the openformats.org community[49]. According to a study in 2012 [108], over 30% of Web pages contain either microformats, microdata, or RDFa. By far the biggest user of RDFa is Facebook with its Open Graph Protocol (OGP) markup using RDFa.

Issues related to URI minting and cool URIs, content negotiation, and redirection are discussed widely on W3C pages[50]. There is some controversy between Schema.org consortium, not using URIs, and the Semantic Web and Linked Data community, advocating their use, but simpler Schema.org descriptions can be transformed into linked data by data conversions. Actually, both Google and Microsoft are creating huge semantic graphs called *Knowledge Graphs* in order to make their searches more semantic. Linked Data movement[51] and technologies are discussed, e.g., in [17, 53]. Lots of datasets are available on the Web not only in the Linked Open Data cloud. For example, the DataHub[52] contains well over 4,000 datasets published using the CKAN data portal platform[53].

SPARQL is a result of several generations of query languages for RDF data. The next version 1.1[54] of the original recommendation[55] in 2008 is about to be finished with several novelties, such as *property paths* describing possible routes through a graph between two graph nodes. Several SPARQL implementations already support such features. More and more CH applications are being built directly on standard (or extended) SPARQL endpoint services, such as the semantic Sampo portal family CultureSampo [72, 99], BookSampo [96, 97], and TravelSampo [98].

---

[46]http://www.w3.org/DesignIssues/LinkedData.html
[47]http://www.w3c.org/RDF/
[48]http://www.microformats.org/
[49]http://www.openformats.org/
[50]http://www.w3.org/TR/cooluris/
[51]http://linkeddata.org/
[52]http://thedatahub.org/
[53]http://ckan.org/
[54]http://www.w3.org/TR/sparql11-query/
[55]http://www.w3.org/TR/rdf-sparql-query/

CHAPTER 4

Metadata Schemas

This chapter concerns metadata, the basis of the Semantic Web, and Linked Data. In the first section different types of metadata are first classified. After this, metadata schemas used for different purposes in representing CH content are presented.

4.1 METADATA TYPES

Metadata [8] means literally "data about data."[1] The American Library Association defined metadata in 1999 in the following way:

> *Metadata is structured, encoded data that describe characteristics of information-bearing entities to aid in the identification, discovery, assessment, and management of the described entities.*

In the context of the Web, the notion of metadata has been given a broader connotation[2]:

> *Metadata is machine understandable information for the Web.*

In the latter sense, the notions of metadata and symbol structures used for *knowledge representation* (KR) [21, 137] become quite closely related. KR is an area of Artificial Intelligence research studying, representing, and inferencing knowledge in terms of symbol structures. Metadata on the Semantic Web is used not only for describing collection items but for modeling the underlying real world and our knowledge about it, too.

Metadata has been traditionally created by information professionals when classifying and indexing collection items during cataloging processes. Nowadays, more and more metadata is produced by non-professional laymen, publishing their contents such as images and videos online. Furthermore, much of the metadata on the Web has been produced automatically using techniques and tools of *data mining* and *information extraction*.

Metadata about collection items can be divided into different types [46].

1. **Administrative metadata,** such as acquisition or location information, is used for managing collections and information sources.

2. **Descriptive metadata,** e.g., the type or material of an artifact, is used for identifying and describing collection contents and related information sources.

---

[1]The Greek word 'meta' means beyond or after and 'datum' is data.
[2]http://www.w3.org/Metadata/

3. **Preservation metadata,** e.g., information about the physical condition of a collection item, is related to preservation management of the collections.

4. **Technical metadata,** such as hardware or software documentation, tells how a system functions or metadata behaves.

5. **Use metadata,** such as rights metadata and search logs, is related to the type and level of use of the collections and information sources.

Metadata can also be divided into indexing and displaying information.

1. **Indexing metadata** is used by search engines and recommender systems for finding stored information.

2. **Displaying metadata** is used to specify, e.g., the particular order in which data should be shown in an application interface.

When using XML schemas, ordering information of elements is fixed by the schema and can be used for displaying purposes. However, when using RDF data, the order of the properties of a resource is not specified. There are ways to specify such ordering information in RDF: 1) RDF contains some specific constructs for representing ordering, such as lists. 2) It is possible to use the ordering of the XML serialization of RDF in a custom way. 3) It is possible to create a custom metaschema in RDF that explicitly specifies property ordering with respect to different resources.

Cultural content in museum collections, libraries, and other content repositories is usually described using *metadata schemas*, also called *annotation schemas*. A metadata schema is a template that specifies the format in which metadata should be represented.

Metadata schemas have been developed for different purposes.

• **Web schemas** for describing documents and objects on the Web.

• **Cataloging schemas** describe a set of obligatory and optional *elements*, i.e., properties by which the metadata for a content item should be described, and constraints for filling in the element values. Metadata element values are filled in when cataloging items. For example, descriptive cataloging metadata may tell the title, author(s), and topic(s) of a book in a library.

• **Harmonization schemas** are ontological models onto which heterogeneous metadata models can be transformed for semantic interoperability. These schemas are not meant to be used for cataloging but can be a source of inspiration for developing such schemas.

• **Harvesting schemas** are used for harvesting heterogeneous metadata from different sources, with the goal of creating an aggregated portal application.

In the following, metadata schemas for CH are discussed along this classification.

4.2   WEB SCHEMAS

The purpose of Web schemas is to provide harmonized metadata elements for describing documents and other objects on the Web. The most notable model for this is Dublin Core and its applications. Elements of Dublin Core and its derivatives are often also used in cataloging schemas, so the distinction between Web schemas and cataloging schemas can sometimes be vague.

4.2.1   DUBLIN CORE

Dublin[3] Core (DC) is a metadata model and framework widely used in libraries and other organizations. DC can be used to describe a wide range of objects, such as books, photos, videos, Web pages, and artworks.

The heart of DC is the DC Metadata Element Set[4] (DCES) that contains 15 standardized[5] elements listed in Table 4.1.

Table 4.1: DC Metadata Element Set				
title	creator	subject	description	publisher
contributor	date	type	format	identifier
source	language	relation	coverage	rights

All DC elements have cardinality $0, ..., n$, i.e., they are optional and can have multiple values (e.g., a book may have several creators). The elements are represented in the order title, creator, subject, etc., as shown on in the table.

The 15 core elements can be extended in an interoperable way by using the "dumb-down" principle. This means that a new "qualified" element can be introduced into the vocabulary with a more specific meaning than an existing element. In this way more specific descriptions can be made interoperable with less specific ones. For example, the element "manufacturer" would be more specific than "creator." This means that when searching for creators manufacturers can also be found. The vocabulary can also be extended by introducing totally new elements. The result of extending and introducing elements in the core is called a *DC application*.

The current DC recommendation is to use an extended version of the core elements called DCMI (Metadata) Terms[6]. It has became one of most popular RDF vocabularies in use, and its specifications are compatible with the ideas of the Semantic Web and Linked Data. DCMI Terms contain the 55 elements of Table 4.2 defined in the namespace `http://dublincore.org/documents/dcmi-terms/`:

This application includes duplicates for all the 15 core elements defined as their refinements. In addition, for example, the core element *date* has been refined with the elements *dateAccepted*, *dateCopyrighted*, and *dateSubmitted*.

---

[3]Dublin refers to Dublin, Ohio, where the headquarters of the OCLC (Online Computer Library Center) organization are situated.
[4]`http://dublincore.org/documents/dces/`
[5]NISO Standard Z39.85-2001 and ISO Standard 15836-2003
[6]`http://dublincore.org/documents/dcmi-terms/`

Table 4.2: DCMI Metadata Terms

abstract	accessRights	accrualMethod	accrualPeriodicity	accrualPolicy
alternative	audience	available	bibliographicCitation	conformsTo
contributor	coverage	created	creator	date
dateAccepted	dateCopyrighted	dateSubmitted	description	educationLevel
extent	format	hasFormat	hasPart	hasVersion
instructionalMethod	isFormatOf	isPartOf	isReferencedBy	isReplacedBy
identifier	isRequiredBy	issued	isVersionOf	language
license	mediator	medium	modified	provenance
publisher	references	relation	replaces	requires
rights	rightsHolder	source	spatial	subject
tableOfContents	temporal	title	type	valid

DCMI Terms specification includes *encoding schemes* that are used to harmonize ways in which element values are expressed. Firstly, there are *vocabulary encoding schemes* that are references to vocabularies, such as Library of Congress Subjects (LCSH) or Medical Subject Heading Headings (MeSH). Secondly, there is also a list of *syntax encoding schemes* mentioned, i.e., references to standard ways of encoding dates, languages, country codes, etc.

There is also a set of *classes* defined and used as range constraints in some elements. For example, the value of *creator* is assumed to be an instance of the class Agent identified by the URI `http://purl.org/dc/terms/Agent`.

Finally, the specification contains a generic *DCMI Type Vocabulary* of classes, such as *Collection*, *Dataset*, *Event*, and *Image*, whose instances can be annotated using the metadata schema.

In addition to technical specifications, the DC community has published various guidelines for using the DC framework, e.g., on how to create application profiles.

4.2.2   VRA CORE CATEGORIES

DC has been used as a basis in more detailed cultural metadata schemas, such as the Visual Resource Association's (VRA) Core Categories[7]. The element set in Core 4 provides elements for the description of 1) works of visual culture, 2) the images that document them, and 3) collections of objects. Relations between works and their images can be indicated. Collections may contain either works or images. Most VRA elements are defined as refinements (subproperties) of DC elements.

An example of an instance of VRA metadata, used in the CHIP portal [5, 7], is given below in RDF Turtle notation:

```
rijks:artefactSK-C-K    vra:type vra:Work ;
  vra:title "The Night Watch" ;   vra:date "1642" ;
  vra:creator ulan:500011051 ;    # Rembrandt
  vra:subject iconclass:45F31 ;   # Call to arms
  vra:culture tgn:7006952 ;       # Amsterdam
  vra:material aat:30015050 .     # Oil paint
```

---

[7]`http://www.vraweb.org/`

Figure 4.1: "The Night Watch" by Rembrandt van Rijn.

The record represents Rembrandt's painting "The Night Watch" in the collections of the Rijksmuseum, Amsterdam (Figure 4.1). The schema has properties such as *vra:type* (the type of the artwork as a reference to the VRA vocabulary), *vra:title* (literal title of the artwork), *vra:creator*, *vra:subject*, *vra:culture*, and *vra:material*. Element values with a namespace are references to underlying domain ontologies[8]: ULAN[9] is an ontology for actors, Iconclass[10] for iconographic descriptions, TGN[11] for historical places, and AAT[12] for concepts of art, architecture, and material culture.

VRA has been endorsed as an extension schema for METS[13] objects that contain images of cultural heritage resources. In the example above, VRA was used as a Web schema but the system is also used as a basis for cataloging systems.

4.3   CATALOGING SCHEMAS

CH collection objects of different types have different characteristics. Their description often requires different metadata models. For example, a painting may depict events that occurred in history or in fantasy, while an artifact, such as a table, is characterized by, e.g., its type, material, and construction. This section presents a few metadata models developed and used for cataloging in libraries, museums, and archives.

---

[8]We will use the term "domain ontology" for domain specific thesauri, controlled vocabularies, classifications, and gazetteers represented in RDF form even if they have not been created as "ontologies."
[9]http://www.getty.edu/research/tools/vocabularies/ulan/about.html
[10]http://www.iconclass.org/
[11]http://www.getty.edu/research/tools/vocabularies/tgn/about.html
[12]http://www.getty.edu/research/tools/vocabularies/aat/about.html
[13]http://www.loc.gov/standards/mets/

4.3.1   CATEGORIES FOR THE DESCRIPTION OF WORKS OF ART (CDWA)

Categories for the Description of Works of Art (CDWA) [9] is a metadata model targeted for cataloging works of art. The CDWA system is more extensive than VRA including 532 categories, i.e., elements and subcategories. However, a small minimal set of *core categories* have been selected for identifying and describing uniquely and unambiguously a work of art or architecture, and there is a Lite version of CDWA available, too. A synopsis of the CDWA core categories divided into groups is given below in order to give a general idea of the CDWA system:

1. OBJECT, ARCHITECTURE, OR GROUP: Catalog Level; Object/Work Type; Classification Term; Title or Name; Measurements Description; Materials and Techniques Description; Creator Description; Creator Identity; Creator Role; Creation Date; Subject Matter; Current Location Repository Name/Geographic Location; Current Repository Numbers;

2. RELATED TEXTUAL REFERENCES AUTHORITY: Brief Citation; Full Citation;

3. CREATOR IDENTIFICATION AUTHORITY: Name; Source; Display Biography; Birth Date; Death Date; Nationality/Culture/Race; Life Roles;

4. PLACE/LOCATION AUTHORITY: Place Name; Source; Place Type; Broader Context;

5. GENERIC CONCEPT AUTHORITY: Term; Source; Broader Context; Scope Note; Source;

6. SUBJECT AUTHORITY: Subject Name; Source; Broader Context.

Rules and examples for using the core subset of the CDWA categories and the VRA Core Categories have been documented in [10].

4.3.2   SPECTRUM

Standard ProcEdures for CollecTions Recording Used in Museums (SPECTRUM)[14] is an XML-based metadata schema for museum collection cataloging. It is an open and freely available collections management standard. The system is originally UK based, but it is now used in some 7,000 museums in 40 countries, and includes translations and localizations.

The standard encompasses the SPECTRUM Schema, the Standards wiki, and SPECTRUM terminologies. It is at the same time both a metadata standard for museum collection documentation and a set of procedural guidelines for cataloging. As in case of CDWA, the schema is extensive, and a simplified version of it called SPECTRUM Essentials has been created for light weight documentation in, e.g., smaller museums.

---

[14]http://www.collectionslink.org.uk/programmes/spectrum

4.3.3   METADATA FORMATS IN LIBRARIES

Libraries have a long tradition of creating metadata records about publications. A widely used standard here has been MARC (MAchine-Readable Cataloging)[15], developed by the Library of Congress in the 60's. This is a sophisticated metadata model for library use. However, the system is hard to learn and use from a human user perspective. For example, the system is based on hard-to-remember three digit tags and subfield codes. Furthermore, MARC pre-dates the Web era and its standards. MARC has therefore been modernized into MARC-XML[16] and there is a more human-friendly derivative of it available, Metadata Object Description Standard (MODS)[17]. The Metadata Authority Description Schema (MADS)[18] is a related XML schema for providing metadata about actors, events, and terms (topics, geographics, genres, etc.).

In many cases documentation of an object contains several descriptions about it, such as metadata about a document, images about it, related reviews, etc. The Metadata Encoding and Transmission Standard (METS)[19] is a "wrapper" standard for representing such collections of metadata descriptions. For the metadata of the collection items, schemas such as DC, MODS, and VRA can be used.

4.3.4   METADATA FORMATS IN ARCHIVES

Archives are the third major type of memory organizations in addition to museums and libraries. Archives, libraries, and museums differ from each other in 1) what they remember and 2) who they serve.

- Archives preserve the source material on which our historical understanding is based; the materials are legal and historical evidence of history. Legal and historical memory require a high degree of user confidence in the authenticity and integrity of records and documents. The materials in archives and manuscript libraries are the unique records created as byproducts of corporate bodies, individuals, and families carrying out their functions, responsibilities, and lives. The documents generated by one corporate body, individual, or family is called a collection or *fond* [116]:

> *A fond is a whole of the documents, regardless of form or medium, organically created and/or accumulated and used by a particular person, family, or corporate body in the course of that creator's activities and functions.*

- Libraries collect individual published books and serials, or bounded sets of individual items. The items in collections are not unique in the sense that there are multiple copies of one

---

[15]http://www.loc.gov/marc/
[16]http://www.loc.gov/standards/marcxml/
[17]http://www.loc.gov/standards/mods/
[18]http://www.loc.gov/standards/mads/
[19]http://www.loc.gov/standards/mets/

publication. Any given copy will generally satisfy the customer's information need equally well.

- Museums differ a lot from each other in terms of the materials they preserve. Even within a single museum, the collections can be quite heterogeneous including, e.g., art, artifacts, books, photographs, etc. The items in a collection are typically unique, such as paintings, but multiple copies may also exist as in libraries.

Museums, libraries, and archives have traditionally served different, though overlapping communities. Museums and libraries have generally served the public and educational and scholarly communities. Many archives, in turn, serve the law, functioning as the institutional memory of specific corporate bodies. Government agencies, public institutions, and businesses have legal requirements pertaining to the keeping of records. These differences in collection materials and in served communities mean that requirements for metadata and cataloging practices in archives are somewhat different from those in museums and libraries. As a result, different metadata formats and standards have been developed and are used in archives.

The main international body in the archiving area is the International Council of Archives (ICA)[20], publishing several archiving standards on its Web site[21]. Other standards include, e.g., the Encoded Archival Description (EAD)[22] used for describing inventories, registers, indexes, and other archiving repositories, and the Encoded Archival Context for Corporate Bodies, Persons, and Families (EAC-CPF)[23] has been developed for representing authorities involved.

4.4   CONCEPTUAL HARMONIZATION SCHEMAS

Schema definitions tackle the problems of *syntactic* and *semantic interoperability* of content object descriptions. Obviously, interoperability problems can be tackled effectively by using a single schema. However, different schemas are needed and used for different kinds of objects in portal applications dealing with cross-domain contents. This section first discusses approaches to schema integration and then presents ontology-based approaches to harmonizing metadata models.

4.4.1   APPROACHES TO SEMANTIC INTEROPERABILITY

Syntactic interoperability can be obtained by harmonizing structural forms for representing data (e.g., by using DC elements or by an XML schema), and by fixing value encoding conventions for element values (e.g., controlled vocabularies, date format, coordinate system, etc.). Semantic interoperability is obtained by shared conventions for interpreting the syntactic representations, e.g., that the property *dc:subject* describes the subject matter of a document as a set of keyword resources taken from an ontology.

---

[20]http://www.ica.org/
[21]http://www.ica.org/10206/standards/standards-list.html
[22]http://www.loc.gov/ead/
[23]http://eac.staatsbibliothek-berlin.de/

Making different metadata schemas semantically interoperable includes two subtasks. Firstly, semantic interoperability of element values has to be addressed using (shared) vocabularies and ontologies. Secondly, interoperability between different schema elements has to be facilitated.

There are two major approaches for making metadata schemas interoperable: Using the dump-down principle is a simple approach for mapping schema elements with each other. Another approach is to create a conceptual ontology model on which different schemas can be transformed and harmonized in this way. Below, Europeana Semantic Elements (ESE) is presented as an example of the former approach. After this, approaches using the latter approach are described: Europeana Data Model (EDM), CIDOC CRM originating from the museum domain, and FRBR from the library domain.

4.4.2   EUROPEANA SEMANTIC ELEMENTS (ESE)

Europeana Semantic Elements (ESE)[24] is an example of a typical DC application. ESE consists of the original 15 DC elements refined with a selection of 21 additional elements from DCMI Terms. In addition, 14 new elements are introduced in the Europeana namespace, such as *country* and *dataProvider*.

ESE is used in the massive European initiative of creating the pan-European CH portal Europeana. The specification is provided as an XML schema for automatic data validation and is normative for Europeana content providers. Data harmonization means that data from different collection systems is transformed into this single schema. ESE is an extension of the DCMI Terms schema and accepts all DC terms including those that cannot be imported into the portal.

4.4.3   EUROPEANA DATA MODEL (EDM)

In spite of its name, ESE and its use in Europeana is not particularly "semantic" from a Linked Data perspective. To address this problem, the Europeana Data Model (EDM)[25] based on RDF has been proposed and is used in the linked open data publication of Europeana contents[26] [51]. The datasets in this publication contain 2.4 million texts, images, videos, and sounds gathered by Europeana, and are freely available in RDF and XML form. ESE is a subset of EDM and can be mapped directly on it.

EDM is a Semantic Web-based framework for representing cross-domain collection metadata in museums, libraries, and archives. The model facilitates richer content descriptions than ESE, and data linking based on shared resources. EDM makes a semantic distinction between the intellectual or technical descriptions of objects that are harvested from the content providers, the object that these descriptions are about, and digital representations of the object. For example, the painting "Mona Lisa" is in the Louvre, but there are numerous copies, photographs, texts, drawings, and statues depicting it in various European museums. Different views of the same object can be represented

---

[24]http://pro.europeana.eu/documents/900548/dc80802e-6efb-4127-a98e-c27c95396d57
[25]http://pro.europeana.eu/web/guest/edm-documentation
[26]http://pro.europeana.eu/linked-open-data

using a special proxy mechanism. It is anticipated that the proxy mechanism will be replaced in the future by named graphs after this model has been standardized.

In contrast to ESE, EDM is not a fixed schema that dictates the way of representing data but rather a conceptual framework or ontology into which more specific models can be attached and interoperability between them enhanced.

4.4.4   CIDOC CONCEPTUAL REFERENCE MODEL (CRM)

Many metadata schemas, such as EDM above, are not intended for cataloging purposes, but are used for data harmonization and integration purposes. A notable example of this is the CIDOC Conceptual Reference Model (CIDOC CRM) [32] (cf. [33]).

The idea here is that different metadata models used for memory organizations for different kinds of collection objects could be transformed into a form conforming to a more foundational semantic ontology model. The model serves as an intellectual guide for format creation, a language for analysis, and a data transportation format.

Dublin Core can be characterized as *document centric* or *object centric*: its elements are used for describing documents or objects. In contrast, CIDOC CRM is *event centric* focusing on the more foundational notion of events related to the creation, use, and maintenance of collection documents and objects. Events can be considered a kind of "semantic glue" that relates different aspects of knowledge (objects, actors, times, places, names, etc.) in a CH collection with each other in a harmonized way. Various mappings and mapping tools for transforming different metadata models into CIDOC CRM, including Dublin Core, LIDO[27], FRBR[28], and EDM[29], have been created and are available[30] for a comprehensive listing with documentation.

CIDOC CRM "provides definitions and a formal structure for describing the implicit and explicit concepts and relationships used in cultural heritage documentation"[31]. This includes not only representation of objects but also their potentially complex provenance information.

The standard specification [28] includes 90 entity classes (numbered by *E1–E90*) and a set of 149 properties (*P1–P149*) relating entities and properties with each other. The core classes of the model as a class hierarchy are presented in Fig. 4.2. Also the properties are organized into a subproperty hierarchy, as in RDFS.

Transformation of collection metadata into this model involves creating an instance of a class in the model. For example, the data record of an artifact can be instantiated based on the class *E22 Man-Made Object*. Instance-class relationships are expressed using the property *P2 has type of*. The object instance is given identifiers and names called *appellations* (*E41 Appellation*), acquisition and ownership information, location information as references to places in an ontology, historical

---

[27]http://www.lido-schema.org/
[28]http://www.ifla.org/VII/s13/frbr/
[29]http://pro.europeana.eu/edm-documentation
[30]See http://www.cidoc-crm.org/crm_mappings.html
[31]http://cidoc.ics.forth.gr/

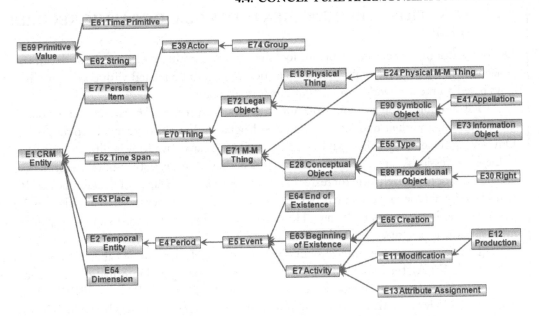

Figure 4.2: Hierarchy of Core Components of CIDOC CRM. Arcs denote *subclass-of* relations. Courtesy of CIDOC CRM community (http://www.cidoc-crm.org/).

information about events related to the object, subject matter descriptions, and physical descriptions of the object (e.g., *P43 has dimension*). The annotation values of the properties are typically references to resources. For example, the values (range) of *P52 has current owner* are actors (*E39 Actor*). During collection transformation, instances for such values have to be created based on collection metadata and/or acquired from shared ontologies.

Much of the contents are described in terms of events (*E5 Event*) and activities of different kinds, such as *E10 Transfer of Custody*, *E11 Modification*, *E65 Creation*, or *E87 Curation Activity*. These are characterized by time (e.g., *P10 falls within* and *P4 has time-span*), place (e.g., *P7 took place at*), and actors involved (e.g., *P11 had participant*).

CIDOC CRM is in use in several organizations and is actively maintained by a community under the auspices of the ICOM CIDOC, the International Committee for Documentation[32] in the museum domain.

[32]http://icom.museum/the-committees/international-committees/international-committee/
international-committee-for-documentation/

4.4.5 FUNCTIONAL REQUIREMENTS FOR BIBLIOGRAPHIC RECORDS (FRBR)

DC originates from the library domain focusing on representing metadata about documents, i.e., collections items in library collections. Collection items can be related with each other in various ways based on the real world processes of intellectual creation.

For example, Homer's epic work Odyssey has been realized in various expressions, i.e., in Greek versions and their translations into other languages. Various other publications are based on Odyssey, e.g., cinematographic adaptations, which in turn can be available in different versions. Each of these "expressions" can be "manifested," i.e., disseminated in a variety of physical formats, such as in hardcover or paperback form for textual expressions, or on VHS tape or DVD for movies. Physical items held by libraries are exemplars of such manifestations. In order to serve librarians and library patrons better in finding and relating CH materials related to creative works and their publications, more versatile conceptual models than DC are needed.

To take these relations into account, the International Federation of Library Associations and Institutions (IFLA), the leading international body in the library and information services domain, develops and standardizes "Functional Requirements for Bibliographic Records (FRBR)"[33], a framework and a family of standards for representing conceptual metadata in the library domain. [93] The term "functional" in these specifications means that the goal of the underlying conceptual work and its implementations is to support four basic task or functions of libraries executed by the customers or the librarians.

1. **Find**. Finding an entity or set of entities corresponding to stated criteria.

2. **Identify**. Identifying an entity (confirm that the entity found corresponds to the entity sought).

3. **Select**. Select an entity that is appropriate to the user's needs.

4. **Obtain** access to the described entity.

The FRBR standard family contains three conceptual entity-relationship models, listed in Table 4.3, each covering an aspect of the data recorded in bibliographic and authority records.

Name	Focus	First Published
FRBR	Functional Requirements for Bibliographic Records	1998
FRAD	Functional Requirements for Authority Data	2009
FRSAD	Functional Requirements for Subject Authority Data	2010

Table 4.3: FRBR Family of Metadata Models

Below, these models are introduced.

---

[33]http://www.ifla.org/VII/s13/frbr/

FRBR

In contrast to document centric DC, FRBR makes a semantic distinction between the entities of *work*, *expression*, *manifestation*, and *item* regarding bibliographical records. Work is a distinct intellectual or artistic creation in an abstract sense, e.g., a story like William Shakespeare's play "A Midsummer Night's Dream." A work is realized through different expressions, i.e., intellectual or artistic realizations of the work. For example, the play "A Midsummer Night's Dream" can be expressed in a critical edition, a bawdlerized version, or a translation. Each expression can be embodied in several manifestations, i.e., in physical embodiments of expressions, such as a particular publication of a given translation. Finally, manifestations are exemplified by items, single exemplars of a manifestation, such as books in library shelves or copies of films in an audiovisual archive. An item can exemplify only one manifestation. This basic model is illustrated in Figure 4.3, where single-headed arrows indicate one-to-one relationships and double-headed arrows one-to-many relationships. In short: using FRBR, it is possible to represent accurately different kinds of items related to an intellectual or artistic creation like "A Midsummer Night's Dream."

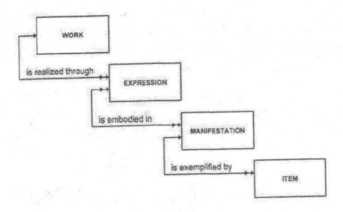

Figure 4.3: Primary entities and relationships in FRBR. Graphics courtesy of [93].

4.4.6 FUNCTIONAL REQUIREMENTS FOR AUTHORITY DATA (FRAD)

Functional Requirements for Authority Data (FRAD) [115], formerly known as Functional Requirements for Authority Records (FRAR), extends the FRBR model above by 1) adding attributes for actors and relationships among actors, 2) by new relationships between actors and their appellations, an 3) telling how information relating to actors is managed by libraries.

4.4.7   FUNCTIONAL REQUIREMENTS FOR SUBJECT AUTHORITY DATA (FRSAD)

The third conceptual model of the FRBR family is Functional Requirements for Subject Authority Data (FRSAD)[34] [160]. FRSAD is a model for the subject relationship between works and their subjects. FRSAD extends the FRBR model with new entities and relationships in the same vein as FRAD. The subject relationship is represented using the entities of Table 4.4 that were already declared in FRBR.

Table 4.4: FRSAD Entities

Entity	Meaning
CONCEPT	An abstract notion or idea
OBJECT	A material thing
EVENT	An action or occurrence
PLACE	A location

The idea is that these entities, as well as the primary entities of FRBR and FRSAD, are used as subject descriptions called *themas* (entity THEMA) that are identified by appellations (entity NOMEN), i.e., by any sign or sequence of signs (alphanumeric characters, symbols, sound, etc.) that a thema is known by, referred to, or addressed as. Figure 4.4 illustrates the relationships between entities in the FRBR family model.

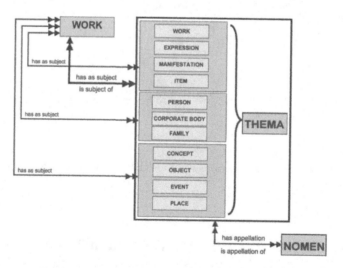

Figure 4.4: FRSAD's relations to FRBR and FRAD. Graphics courtesy of [160].

---

[34]http://www.ifla.org/node/5849

4.4.8 FRBROO

Standardization work on FRBR in the library domain and on CIDOC CRM in the museum domain share the same goal: making information more findable by harmonizing heterogeneous formats in use. Furthermore, both works are based on similar kinds of semantic conceptual entity-relationship models. A natural next step forward is to apply the same approach to harmonizing FRBR and CIDOC CRM themselves. For this purpose, an international working group was formed in 2003. The resulting model, *FRBR object oriented (FRBRoo)* [12], is an extension to CIDOC CRM, where FRBR entities are added into the hierarchical entity structure of CIDOC CRM (cf. Table 4.2), and relationships and attributes used in the models are aligned. FRBRoo adds into FRBR the event-centric aspects of CIDOC CRM, and a number of other refinements. The goal of FRBRoo is to create an ontology model for harmonizing bibliographical metadata from libraries with collection metadata from museums. Aligning EDM with FRBRoo is being planned.

4.5 HARVESTING SCHEMAS: LIDO

Metadata formats are also used for data harvesting purposes, providing a common model onto which different publishers could transform their databases. This section presents a recent development toward this: Light Weight Information Describing Objects (LIDO)[35]. LIDO is targeted to be used, e.g., in the linked data-based versions of the Europeana portal.

LIDO is an XML harvesting schema. It is intended for delivering metadata for aggregated online services such as portals; it is not intended to be used as a basis for a collection management system or to support loan and acquisition activities. As an aggregation format, the model can be used for describing different kinds of objects, including art, architecture, cultural history, history of technology, and natural history. The metadata may originate from different collection management systems using different metadata models in different languages. For this purpose, LIDO combines features of the U.S.-based CDWA Lite, the German *museumdat* schema[36], and the system is aligned with the British SPECTRUM collections management standard. The system is compliant also with the CIDOC CRM standard. Below, the structure of LIDO is presented based on the standard documentation [27].

A LIDO record may include 14 groups of descriptive and administrative elements of metadata, three of which are mandatory (cf. Table 4.5). The element Subject Set is used for describing the subject matter of the object using the categories of Table 4.6. From a semantic perspective, a particularly versatile element in LIDO is *Event Set*. It describes the object in terms of event instances that are described using the event element categories of Table 4.7. Events occur in time, in different places, and are associated with actors and other resources. This makes it possible to express detailed indirect semantic relations between collection items and other resources in space and time. Benefits of event centric modeling are discussed in some more detail in end of this chapter.

[35]http://www.lido-schema.org/
[36]http://www.museumdat.org/

Table 4.5: LIDO metadata elements

Metadata Type	Elements
Object Classifications	Object/Work type (mandatory)
	Classification
Identification	Title/Name (mandatory)
	Inscriptions
	Repository/Location
	State/Edition
	Object Description
	Measurements
Events	Event Set
Relations	Subject Set
	Related Works
Administrative	Rights
	Record (mandatory)
	Resource

Table 4.6: LIDO Subject Set Elements

Category	Meaning
Extent Subject	When there are multiple subjects, a term indicating the part of the object/work to which these subject terms apply.
Subject Concept	Provides references to concepts related to the subject of the described object/work.
Subject Actor	A person, group, or institution depicted in or by an object/work, or what it is about, provided as display and index elements.
Subject Date	A time specification depicted in or by an object/work, or what it is about, provided as display and index elements.
Subject Place	A place depicted in or by an object/work, or what it is about, provided as display and index elements.
Subject Event	An event depicted in or by an object/work, or what it is about, provided as display and index elements.
Subject Object	An object, e.g., a building or a work of art depicted in or by an object/work, or what it is about, provided as display and index elements.

4.6   HARVESTING AND SEARCHING PROTOCOLS

In the distributed content creation model of this book, local memory organizations either 1) open an API for the centralized portal to search for contents (in federated search) or 2) let the portal to harvest metadata published by the local producers at some Web address. This section introduces protocols for metadata harvesting in legacy CH applications, starting form traditional protocols in use, and ending up with the linked data solution approach: SPARQL endpoints.

Table 4.7: LIDO Event Description Elements

Category	Meaning
Event ID	Event identifier.
Event Type	The nature of the event associated with an object/work.
Role in Event	The role played within this event by the described entity.
Event Name	An appellation for the event, e.g., a title, identifying phrase, or name given to it.
Event Actor	Wrapper for display and index elements for an actor with role information (participating or being present in the event).
Culture	Cultural context.
Event Date	Date specification of the event.
Period Name	A period in which the event happened.
Event Place	Place specification of the event.
Event Method	The method by which the event is carried out.
Materials/Technique	Indicates the substances or materials used within the event (e.g., the creation of an object/work), as well as any implements, production or manufacturing techniques, processes, or methods incorporated.
Thing Present	References another object that was present at this same event.
Event Related	Display and index elements for the event related to the event being recorded.
Event Description	Wrapper for a description of the event, including description identifier, descriptive note of the event, and its sources.

4.6.1   SEARCHING WITH Z39.50, SRU/SRW, AND OPENSEARCH

Several protocols have been developed for federated search. Z39.50 is a client-server protocol for searching and retrieving information from remote databases. The standard is maintained by the Library of Congress and has been standardized by the ANSI/NISO standard Z39.50 and ISO 23950. This protocol has been widely used in legacy CH applications. However, it was created already in the pre-WWW era, is not based on HTTP, and is dated.

Z39.50 has been modernized into the Search/Retrieval via URL (SRU)[37] protocol that uses the HTTP protocol and REST. SRU has a twin protocol called Search/Retrieve Web Service (SWR) that is based on Web Service SOAP messages. Search queries in SRU and SRW are expressed using the simple Contextual Query Language (CQL)[38], a standard based on Z39.50. The result set is returned as an XML document. Any XML serialization of database objects can be used for representing results, e.g., the XML based Library of Congress Standards MARC-XML[39], MODS[40], or MADS[41].

OpenSearch[42] protocol is an attempt to standardize search engine APIs. It specifies formats for expressing queries to search engines and the responses returned. This facilitates querying of search engines in a harmonized way, and syndication of search results in federated search across the Web.

---

[37]http://www.loc.gov/standards/sru/specs/index.html
[38]Formerly known as Common Query Language.
[39]http://www.loc.gov/standards/marcxml/
[40]Metadata Object Description Schema, http://www.loc.gov/standards/mods/
[41]The Metadata Authority Description Schema, http://www.loc.gov/standards/mads/
[42]http://www.opensearch.org/

4.6.2 HARVESTING WITH OAI-PMH

A widely used system targeted for only harvesting metadata is the Open Archives Initiative Protocol for Metadata Harvesting (OAI-PMH)[43]. The OAI-PMH protocol is based on HTTP where request arguments are issued as GET or POST parameters of a URL. There are six data request types available called "verbs," such as *ListRecords* for creating and fetching a list of records. Dublin Core is the minimal format specified for basic interoperability in querying. For example, the following request would return the records of a data repository from the given timestamp 2002-11-01 on.

```
http://archive.org?verb=ListRecords&from=2002-11-01
```

Responses are encoded in XML syntax. OAI-PMH supports harvesting records in any metadata format encoded in XML.

Harvesting protocols make it possible to constrain the amount of data harvested, e.g., to only records that have been changed or added in a repository after the previous harvest. It is of course also possible to harvest every time the whole local dataset by simply downloading it from a URL location, and then extract out changes at the service side of the portal system. This model is simpler for the local content producer because there is no need for implementing and maintaining an OAI-PMH server. However, downloading large databases is often not feasible.

4.6.3 SPARQL ENDPOINT FOR LINKED DATA

The standard way of making federated queries to distributed linked data repositories is to use SPARQL endpoints, based on the SPARQL Protocol for RDF (SPROT)[44], a query language and protocol for RDF (cf. Section 3.2.2). A SPARQL endpoint enables machine and human end-users to make SPARQL queries to an RDF repository in an easy way by using HTTP. The results for a query are returned in one or more machine-processable formats, typically in XML, RDF, or JSON.

For example, assume a library has set up a SPARQL endpoint at the URL

```
http://www.my-library.org/sparql/
```

serving metadata of a collection of books annotated using Dublin Core. The goal is to find all books published in 2012 with their authors. In SPARQL this amounts to selecting (SELECT clause) values for query variables, say ?book and ?author, in such a way that the query constraint expressed as a graph pattern (WHERE clause) using the variables can be satisfied against the triplestore. In this case, the following SPARQL query can be used:

```
PREFIX dc: <http://purl.org/dc/elements/1.1/> SELECT ?book ?author
WHERE { ?book dc:creator ?creator .          ?creator :name ?author .
        ?book dc:time 2012 . }
```

---

[43]http://www.openarchives.org/
[44]W3C recommendation since 2008, http://www.w3.org/TR/rdf-sparql-protocol/

The HTTP call for executing this is

```
GET /sparql/?query=...encoded query here... HTTP/1.1
Host: www.my-library.org
User-agent: my-sparql-client/0.1
```

The results can then be returned as a variable binding list using XML:

```
HTTP/1.1 200 OK Date: Fri, 06 Aug 2012 14:55:12 GMT
Server: Apache/1.3.29 (Unix) PHP/4.3.4 DAV/1.0.3 Connection: close
Content-Type: application/sparql-results+xml

<?xml version="1.0?''>
<sparql xmlns="http://www.w3.org/2005/sparql-results#">

 <head>     <variable name="book"/>
   <variable name="author"/> </head>
 <results distinct="false" ordered="false">      <result>
    <binding
name="book"><uri>http://www.my-library/book/b3215</uri></binding>
     <binding name="author"><literal> John Doe</literal></binding>
   </result>      <result>
     <binding
name="book"><uri>http://www.my-library/book/b1415</uri></binding>
     <binding name="author"><literal> Susan Smith</literal></binding>
   </result>     ...
</sparql>
```

Here $<head>$ markup identifies the query variables and each $<result>$ contains a binding for the variables that satisfies the query pattern. In this case, URIs of books with related author names as literal values are returned. In the underlying metadata model, each book resource has a `dc:creator` property with an author resource as a value, and each author resource has a `:name` property with a literal value indicating the author's name.

The formulation of the queries and the human-readable presentation of the results are typically created by a software agent, not a human user. However, SPARQL endpoints are also used by humans for inspecting datasets and for filtering out subsets of the data. In this case, a Web browser and a URL address with a SPARQL query as an HTTP GET parameter in the query sting can be used easily.

A SPARQL endpoint can be used not only for searching results in federated search but also for harvesting linked data. However, using the endpoint for harvesting large sets of data by querying slows the service down for other users. Downloading files is then more feasible.

4.7   DISCUSSION: OBJECT, EVENT, AND PROCESS MODELS

DC and its application profiles are examples of *object-centric metadata models* where information is described in terms of features assigned directly to objects. In Dublin Core Metadata Element Set, the values of the features are represented by literals, while in DCMI Metadata Terms also resources are also used (instances of classes) in many cases, making the model compatible with the Semantic Web and Linked Data. If only literal values are used and not URI references, the data cannot be regarded as linked.

From a knowledge representation viewpoint the object-centric approach of DC has some fundamental limitations. Consider, for example, the creation, use, and collection of a CH object, say the painting "Mona Lisa," through history. Its metadata description involves different dates, places, and actors, e.g., that "Mona Lisa" was created by Leonardo da Vinci in Florence in 1503–1506, the painting was used by various private owners in different places and times (e.g., by Gian Giacomo Caprotti da Oreno (Salai) after Leonardo's death), was acquired by King Francois I of France, was stored in several collections (e.g., Palace of Fontainebleau and Palace of Versailles) before its current placement in the Louvre.

Representing how the painting is related to places in an object-centric way with direct metadata elements could be done using the DCMI Term dcterm:subject for subject matter, and by introducing subelements of *dcterm:spatial*, such as *placeOfCreation*, *placeOfUsage*, *placeOfAcquisition*, and *placeOfExhibition*. However, such "flat" object-centric representations lead to several problems when aiming at more accurate semantic representations of CH:

- Rich semantics gets embedded implicitly in the properties and is not explicit for the machines to reason, which complicates inferencing in intelligent applications.

- The number of properties in the model easily explodes, making the model complicated and causing problems when aggregating and harmonizing heterogeneous contents. For example, representing the acquisition of an object would require direct properties such as *dateOfAcquisition*, *placeOfAcquisition*, and *actorOfAcquisition* so that the time, place, and actor of the event could be represented without confusing them with times, places, and actors of other events (creation, usage, etc.) related to the object. If the painting is acquired several times, then the times, places, and actors of the different acquisition events would be mixed even using this approach.

An approach to address these issues is to use *event-centric metadata* models. The idea here is to describe objects (partly) in terms of *events* in which they have been involved, such as the creation of "Mona Lisa" in Florence in 1503–1506, its use by Louis XIV, acquisition of the painting to the Louvre, and its exhibitions in museums. By representing events as independent resources of their own, each event can be associated with its particular place, date, and actors involved, and the relations of the object to different places, dates, and actors are established in an accurate way via events. At the same time, the need for introducing specialized, semantically complex metadata elements for the relations is diminished, leading to a simpler model. Finally, events can be shared by different annotations, which can be used as the basis for interlinking and harmonizing content, and for sharing work in creating shared historical event gazetteers. For example, historical events such as the French Revolution or World War II are frequently referred to in related CH materials.

The idea of event-centric modeling and cataloging can be developed even one step further into the notion of *process-centric metadata models* and cataloging [85]. Here the focus is on describing and storing knowledge about intangible CH processes, such as traditional skills (e.g., handicraft or dancing), cultural processes and practices (e.g., farming or political decision making), or narrative stories (e.g., flow of history or traditional epics). These kinds of CH phenomena involve events that follow each other in time and space under some ordering constraints, and decompose into a series of subevents or tasks.

By using events and their narrative structures with causal relations, it is possible to address the famous "five Ws and one H questions" of journalism that are often regarded as a basis in information gathering (cf. Table 4.8).

Table 4.8: Five Ws and one H
Who is it about?
What happened?
When did it take place?
Where did it take place?
Why did it happen?
How did it happen?

4.8   BIBLIOGRAPHICAL AND HISTORICAL NOTES

An introduction to metadata is presented in [8]. More details and documentation about the specific metadata models presented can be found through their home pages indicated in the text.

A particularly significant metadata model for Linked Data has been Dublin Core, supported by a community that maintains Web pages documenting the DC system, its use, and ongoing developments[45]. The community has organized since 1995 an annual metadata workshop that was changed in 2001 into the conference series International Conference on Dublin Core and Metadata Applications[46] with lots of papers related not only to DC but to issues of metadata and knowledge organization systems more generally.

Metadata schemas specify data formats but do not tell how to fill the element values in the formats. Additional standards are necessary for 1) the format or choice of element data values as well as 2) for cataloging practices, i.e., for the description of data content. An example of a data content standard is the Cataloging Cultural Objects (CCO) guidelines[47]. In the CH domain, data value standards have been traditionally specified by classification systems and controlled vocabularies/thesauri.

---

[45] http://dublincore.org
[46] http://dublincore.org/workshops/
[47] http://www.vraweb.org/ccoweb/cco/index.html

The FRBR model is separate from any particular cataloging standard in use, such as Anglo-American Cataloguing Rules (AACR2[48]) and International Standard Bibliographic Description (ISBD).

However, AACR2 has been succeeded by a new standard for cataloging: Resource Description and Access (RDA)[49] that is based on FRBR. RDA is being developed by a consortium of major national library organizations that are planning to adapt the system in their processes. Another significant development of FRBR is its harmonization in FRBRoo with the museum standard CIDOC CRM.

Ontologies and the Semantic Web have made their way to metadata models. Collaboration for solving interoperability issues between different communities developing their own metadata standards for CH content is getting deeper, which is a promising sign from a Linked Data perspective.

---

[48]AARC2 stands for the second edition of AARC.
[49]http://www.rda-jsc.org/index.html

CHAPTER 5

Domain Vocabularies and Ontologies

There is a long tradition in memory organizations to use classifications, controlled vocabularies, and authority files for harmonizing data indexing, for organizing collections, and for information retrieval. Domain ontologies is a step forward in this tradition, facilitating more accurate metadata descriptions that can be utilized also by the machines in harmonizing metadata descriptions, fostering interoperability across different organizations and domains, and in data linking.

In this chapter the ambiguous notions of "ontology" and "vocabulary" are first introduced. Then different types of domain ontologies are discussed from a CH perspective and Semantic Web standards for representing ontologies are reviewed. After this, ontology types needed in CH applications are explained in more detail.

5.1 APPROACHES TO ONTOLOGIES

The term *ontology* is ambiguous with different meanings in Philosophy, Lexicography and Linguistics, Terminology research, Information and Library Sciences, and Computer Science. Furthermore, the term is mixed with the term *vocabulary* that has its own established connotations. According to W3C[1] "vocabularies define the concepts and relationships (also referred to as 'terms') used to describe and represent an area of concern." This means that there is no clear distinction between "ontologies" and "vocabularies" in the context of the Semantic Web, even though the term "ontology" is usually preferred for more complex and formal knowledge organization systems. Furthermore, both terms are used to refer to different kinds of knowledge organization systems, such as large domain-specific gazetteers, classifications, and controlled vocabularies of up to millions of concepts, or small domain independent metadata schemas of tens of concepts.

Below, perspectives to ontology in different scientific disciplines are shortly outlined.

5.1.1 PHILOSOPHY

The original meaning of the term ontology comes from Philosophy: Ontology (with a capital letter) is a branch of Philosophy focusing on studying the existence, structure, and nature of things as they are. This notion originates from Aristotle's (384–322 BC) ideas of Metaphysics. In his ontology, Aristotle identified the ten categories describing the world listed in Table 5.1.

[1]http://www.w3.org/standards/semanticweb/ontology

Category	Example
Substance	A chair
Quality	The dress is red
Quantity	The size of the painting is 30x40cm
Relation	The bridge is half the length of the Golden Gate
Where	The Louvre is in Paris
When	Columbus sailed to America in 1492
Position	The statue is standing
Having	John Doe owns this vase
Action	The book was published in 2012
Passion	The portal solicits more content providers

Table 5.1: Ten ontological categories of Aristotle

Porphyry of Tyre (234–c.305), a Greek Neoplatonist, arranged ontological categories into a dichotomy hierarchy of sub/supertypes, thus introducing the idea of a semantic net. For example, *body* can be *animate* or *inanimate*, *animate* can be *rational* or *irrational*, and so on. Medieval scholastics and logicians, such as Peter of Spain, developed ontological ideas further, and finally "Ontology" as a named discipline emerged in the 17th century. Today, ontology research in Philosophy has a strong theoretical bias with the aim of developing formal models of foundational categories and logic behind everything [137].

5.1.2 LEXICOGRAPHY AND LINGUISTICS

In linguistics, the idea of ontologies is closely related to the idea of semantic dictionaries, where words are organized not alphabetically but according to their meaning. The design, compilation, use, and evaluation of dictionaries is studied in the field of *lexicography*.

An early example of a semantically organized dictionary is Roget's Thesaurus [86] that has been edited since its first publication in 1852. Here the world is divided into 1,000 categories organized into a hierarchy. The top level consists of six categories:

```
1. Abstract relations 2. Space
3. Matter 4. Intellect
5. Volition 6. Affections
```

The next levels in *2. Space*, for example, are classified as follows:

```
CLASS 2. Space    Space in general
   I Abstract space
      180 Indefinite space {Noun: space, extension, extent, expanse,...
                            Verb: reach, extend,...
                            Adj: spacious, roomy,...
                            Adv: extensively, ...}
      181 Definite region ...
      182 Limited space ...    II Relative space
      183 Situation ...    ...
```

The final categories (180, 181,... in the example) list their nouns, verbs, adjectives, and adverbs. This kind of organization is useful, for example, when looking for alternative ways of expressing thoughts.

WordNet [38] is arguably the most widely used (psycho-)linguistic vocabulary/ontology on the Semantic Web. It is part of the LOD cloud with many mappings to related datasets. The English WordNet has been translated at least partly into many other languages[2].

Word meanings in WordNet are organized into cognitive synonym sets called *synsets* that refer to concepts. Homonymous word forms with different meanings belong to several synsets. The synsets based on substantives, verbs, adjectives, and adverbs are organized into subnets of their own.

Substantives are organized into a *hyponym/hyperonym*[3] hierarchy that corresponds to a subclass of hierarchy. A distinction is made between classes (common nouns) and instances (individual persons, places, etc.) as in RDF. All noun hierarchies ultimately go up to the root node "entity." The hyponymy relation is transitive: if an armchair is a kind of chair, and if a chair is a kind of furniture, then an armchair is a kind of furniture. Some synsets are linked together with part-whole relations.

Also, verb synsets are organized into hierarchies where more specific verb meanings are called *troponyms*. For example, "whisper" is a troponym of "talk" that is a troponym of "communicate." Verb synsets entailing each other are linked with each other, e.g., "try" with "succeed."

For organizing adjective synsets, the antonym relation is used. For example, "wet" is associated with "dry." The subnet of adverbs is small because most adverbs in English are derivatives of adjectives.

Most relations in WordNet link words of the same part of speech together, but there are also a few morphosemantic links in use associating, e.g., locations and actions. For example, the noun synset "sleeping_car" is associated with the verb synset "sleep."

The English WordNet contains 117,000 synsets and includes many glosses explaining the concepts. It is freely available also in RDF form and in Prolog for logic programming[4].

There are also large linguistic semantic databases available based on the idea of representing words and their meanings in terms of *semantic roles* related to the words. FrameNet[5] is a lexical database of English that is both human- and machine-readable. It is based on annotating examples of how word senses are used in actual texts. The underlying data model is based in *frame semantics*, where events, relations, and entities are described in terms of their participants, such as *Creator, Beneficiary, Manner, Place, Time*, etc. FrameNet contains over 10,000 word senses. Proposition Bank (PropBank)[6] is another related project with a large corpus annotated with semantic roles. In VerbNet[7] over 6,000 verbs are described in terms of their thematic roles, selectional restrictions on the arguments, and frames. These kinds data resources can be used, e.g., when extracting semantic structures, such as roles and events, in texts.

---

[2]http://www.globalwordnet.org/
[3]Hyponym means more specific and hyperonym more generic term.
[4]http://www.w3.org/2006/03/wn/wn20/
[5]https://framenet.icsi.berkeley.edu/fndrupal/home
[6]http://verbs.colorado.edu/~mpalmer/projects/ace.html
[7]http://verbs.colorado.edu/verb-index/

5.1.3 TERMINOLOGY

Terminology is the study of terms and their use. The major problem addressed here is that words are used in every day life and in different professional contexts with different meanings leading to confusion. Terminological analysis is used to clarify language use by creating specific terminologies for different communities with precise, harmonized definitions. The analysis may be multi-lingual. Terminological definitions are typically *normative*, i.e., they are recommendations for a particular interpretation and use of terms in a domain.

Terminological analysis is a standardized methodology (by ISO 704:2009) for defining concepts [143]. It is based on extending the Odgen-Richards triangle[8] with the notion of definition into a tetrahedron that has following related notions in its corners: Object (the real world object), Idea (the underlying abstract concept), Term (word used to refer to the idea), and Definition (statement describing the idea).

Major sources of ambiguity in determining the references for a word include the following.

- **Synonymy,** where several words have the same meaning, e.g., "buy" and "purchase."

- **Polysemy,** where a single word has different, but related meanings, e.g., "head" of an arrow vs. "head" of a man.

- **Homonymy,** where a single word has different unrelated meanings, e.g., "bank" as a business and as a river bank.

The tool for analyzing and defining concepts is to organize them into a *concept system* that is represented as a graph using four major relation types:

1. **Equivalence** relations,

2. **Generic** relations (hyponymy),

3. **Partitive** relations (part-of), and

4. **Associative** relations (remaining other relations).

Terminology differs from lexicography in studying concepts, conceptual systems, and their labels (terms), whereas lexicography studies words and their meanings.

5.1.4 INFORMATION AND LIBRARY SCIENCE

In Information Science, Library Science, and Information Technology thesaurus construction is studied [1]. Here the goal is to create controlled vocabularies for indexing or tagging purposes and for use in information retrieval systems. The basic use pattern is that during indexing a cataloger selects thesaurus terms for metadata element values describing the content in a harmonized way. During

---

[8]A famous simple model illustrating that linguistic symbols (words) are related to corresponding references of thought (ideas) and the objects they represent [113].

information retrieval the same terms can be used for constructing queries. Lots of standards have been created for thesaurus construction. In the following brief overview, the widely used monolingual standard ISO 2788, its British equivalent BS 5723, and the U.S. Standard ANSI/NISO Z39.190-1993 are used by default.

A thesaurus consists of terms and semantic associations between them. The main categories of them are equivalence, hierarchical, and associative relations, as shown in Table 5.2. However, there are many more refined relations that can be employed for partial equivalence, class-instance relationship, refined hierarchical relationships, of refined associative relationships, as in terminology.

Table 5.2: Standard semantic relations of a thesaurus

Category	Abbreviation	Meaning
definition	SN	Scope note, textual description
equivalence	UF	Use for, prefix indicating non-preferred term
	USE	Prefix indicating preferred term
hierarchical	BT	Broader term
	NT	Narrower term
associative	RT	Related term

An example of a few thesaurus terms is presented below where preferred terms are capitalized:

```
CATERING EQUIPMENT    UF Kitchen equipment

Kitchen equipment    USE CATERING EQUIPMENT

COOKING APPLIANCES    SN Cooking ware, tableware etc.
  BT CATERING EQUIPMENT    BT DOMESTIC APPLIANCES
  NT COOKERS    NT OVENS

DOMESTIC APPLIANCES    SN Appliances used at home
  NT COOKING APPLIANCES    ...
...
```

When indexing, terms taken from a thesaurus are usually considered independent from each other. However, it is also possible to combine terms in order to create more complex concepts by *coordination*. Then several terms making a combined concept are explicitly combined using a given syntax. For example, *Furniture* and *Conservation* could be combined into *Furniture—Conservation*.

There are two kinds of coordination differing in when coordination is done in relation to an information retrieval event:

- **Pre-coordination** is done before information retrieval by a thesaurus developer or content indexer.

- **Post-coordination** is performed during information retrieval when forming and processing queries.

Part-of relations used in thesauri and other knowledge organization systems can be refined further when more accurate analysis is needed. For example, in Table 5.3 a set of more refined partitive relations are listed.

Table 5.3: Different part-of relations	
Relation	**Example**
part ⇆ whole	keyboard ⇆ computer
member ⇆ set	tree ⇆ forest
piece ⇆ whole	frame ⇆ painting
material ⇆ object	silver ⇆ spoon
phase ⇆ process	childhood ⇆ life
place ⇆ region	Helsinki ⇆ Finland

In the same vein, associative relations can be analyzed in more detail. In this case, the number of different relations easily explodes, because there can be so many different associative relations between concepts, as illustrated in Table 5.4.

Table 5.4: Different associative relations	
Relation	**Example**
cause ⇆ effect	war ⇆ death
producer ⇆ product	artist ⇆ painting
activity ⇆ actor	weaving ⇆ artisan
activity ⇆ location	nesting ⇆ tree
object ⇆ location	chimney ⇆ roof
object ⇆ activity	apple tree ⇆ fruit gathering
tool ⇆ function	hammer ⇆ nailing
...	...

5.1.5   COMPUTER SCIENCE

In Computer Science, the term ontology (with a lower case letter) is used to refer to a formal data structure that can be processed using algorithms [48]:

> *An ontology is a formal, explicit specification of a shared conceptualization.*

The keywords in the definition are:

- *formal*, i.e., an ontology has well-defined syntax and semantics,

- *explicit*, i.e., an ontology can be represented and processed algorithmically,

- *shared*, i.e., an ontology is agreed upon in a community and facilitates communication between its member agents, and

- *conceptualization*, i.e., an ontology presents a model of the real wold.

5.2   SEMANTIC WEB ONTOLOGY LANGUAGES

In Semantic Web and Linked Data the key standard models for representing ontologies are RDF Schema, Simple Knowledge Organization System SKOS, and Web Ontology Language OWL. Below, a brief overview to these standards is given, with reference for further readings on the topics.

5.2.1   RDF SCHEMA

RDFS introduces object-oriented modeling into RDF. The idea is to describe a domain of discourse in terms of classes, individual instances belonging to the classes, and properties that describe classes and individuals. A class, say *Painting*[9], represents the set of its instances (e.g., Mona Lisa) that share the general properties of the class. These properties are specified when an instance is created, i.e., the class is *instantiated*. For example, if *Painting* has properties *dc:title* and *dc:creator*, then the string "Mona Lisa" may be set for the title and a URI reference to Leonardo Da Vinci for the creator when instantiating Mona Lisa. Instance-class relationship is represented by the `rdf:type`[10] property arc from the instance (URI) to its class (URI).

Classes are organized into a subclass hierarchy using the property `rdfs:subClassOf`. An instance is automatically considered to belong to not only the class where the `rdf:type` arc points to, but also to its superclasses, inheriting their definitions, too. Also, the properties are organized into a hierarchy based on the property `rdfs:subPropertyOf`. Properties in RDFS are resources and can therefore have properties of their own.

RDFS also introduces into RDF the idea of property constraints: 1) A *range constraint* tells that the value of a property (e.g., *dc:creator*) must always be an instance of a specific class (e.g., *Person*). 2) A *domain constraint* tells in the same way that the property subject must be an instance of a specific class, e.g., that the property *dc:creator* can be used only for the instances of the class *Work*.

An RDFS ontology therefore consists of 1) a class hierarchy, 2) a property hierarchy, and 3) a set of property constraints. A related RDF dataset can be created by instantiating the classes with specific property values.

5.2.2   SIMPLE KNOWLEDGE ORGANIZATION SYSTEM SKOS

The main goal of the SKOS standard is to provide a light-weight ontology format in RDF for representing vocabularies, such as legacy thesauri and classifications in use. From a philosophical perspective, the focus of modeling is therefore rather to describe terms and thesaurus structures to present a conceptualization of the real world, even if thesauri are used for representing concepts of the real world.

The key concept in SKOS is the class *skos:Concept*. A SKOS vocabulary is built as a semantic network connecting instances of this class that represent terms of a thesaurus or a classification, or collections of concepts (*skos:Collection*, *skos:OrderedCollection*). The vocabulary includes pre-defined properties for terminological equivalence (e.g., *skos:prefLabel* and *skos:altLabel* for preferred and

[9]By convention, class names usually start with a capital letter.
[10]By convention, property names usually start with a lower case letter.

alternative concept labeling), relations for representing thesaurus hierarchies, such as *skos:narrower* and *skos:broader*, and relations for concept associations, such as *skos:related*. Each SKOS vocabulary is an instance of the class *skos:ConceptScheme*.

SKOS specification includes a set of *integrity conditions* that can be used for validating concept schemas, and a set of logic rules that can be used for enriching vocabulary relations by reasoning. The model is compatible with RDF(S) and the more versatile OWL standard for representing ontologies, and can be extended using their constructs.

SKOS is "simple" yet versatile enough for capturing much of the semantics of existing thesauri. It is widely used in Linked Data, focusing on using light-weight ontologies and RDF data based on existing data resources.

5.2.3   WEB ONTOLOGY LANGUAGE OWL

OWL was created for addressing semantic limitations of RDFS, such as the following.

- Constraints of a property cannot be specified class-wise in RDFS, e.g., that the range of *dc:creator* for books should be different (e.g., *Author*) from that of a symphony (e.g., *Composer*).

- Cardinality of properties cannot be specified, e.g., that a *Person* has two parents.

- Basic semantic properties of relationships are often needed in ontology modeling and reasoning but are missing in RDFS. For example, there are inverse properties (e.g., wife-of vs. husband-of), transitive properties (e.g., ancestors of a person are ancestors of her children, too.), functional relations (e.g., a person has a unique mother and father), and inverse-functional properties (e.g., a social security number uniquely determines its holder).

- Defining classes using basic set operations is not possible in RDFS. For example, it is not possible to say that the class *Person* is the union of *Man* and *Woman*.

OWL extends RDFS with new constructs for modeling such ontological situations. By increasing expressive power, more accurate definitions can be created and more reasoning performed, but at the price of computational efficiency. To balance this expressiveness-efficiency trade-off, three versions of OWL were standardized in 2004[11]: OWL Lite, OWL DL, and OWL Full, in the order of increasing expressive power. In 2009, the next OWL specification, called OWL 2[12], was standardized with three *language profiles* for different use cases:

- OWL 2 EL for ontologies with a large number of classes/properties,

- OWL 2 QL for ontologies with lots of instance data, and

- OWL 2 RL for applications that need scalable reasoning and decent expressive power at the same time.

---

[11]http://www.w3.org/2004/OWL/
[12]http://www.w3.org/TR/owl2-overview/

OWL 2 also employs an alternative "Manchester syntax" for specifying ontologies in a concise, human friendly notation. OWL 2 is backward compatible with the original OWL specification.

A challenge of OWL from a practical perspective is conceptual complexity. More expressive ontological constructs are harder to master by human users and cannot often be created automatically from existing data sources. Another difficulty in many cases is that many complex real world concepts are hard to model in terms of precise logical formulations. For example, concepts may be vague by their nature, or our knowledge about them may be uncertain or incomplete, leading to problems of non-monotonicity in reasoning. As a result, complex ontological modeling is usually not used in Linked Data where practical applications are developed using large existing datasets and light-weight ontologies.

5.3   ONTOLOGY TYPES

The notion of *knowledge organization systems* (KOS) encompasses all schemes for organizing information, including classification and categorization systems, thesauri, subject headings, authority files, gazetteers, term lists, dictionaries, and ontologies. This section discusses first the differences between classifications, thesauri, and ontologies. After this, different types of ontologies used in semantic CH systems are introduced.

5.3.1   CLASSIFICATIONS, THESAURI, AND ONTOLOGIES

The idea of *classification systems* is to organize and group items together into categories. Classification systems have been developed for various fields of science, such as biology, medicine, and library and information sciences.

There are three major approaches in use.

- *Enumerative subject headings* are simple lists of classifying categories.

- In *hierarchical classifications*, topics higher in the hierarchy cover in content their subtopics lower in the hierarchy.

- In *faceted classification*, documents are classified at the same time along several orthogonal classifications. Faceted classifications can be used as a basis for faceted search and are widely used in semantic portals for CH applications [63, 132].

Classifications can be classified along their topic area: universal classification schemes, such as the Dewey Decimal Classification (DDC)[13], the world's most widely used classification, the multilingual Universal Decimal Classification (UDC)[14], and the Library of Congress Classification (LCC)[15] cover all aspects of life, while there are also classification systems for special areas, such as

---

[13]http://www.oclc.org/dewey/default.htm
[14]http://www.udcc.org/
[15]http://www.loc.gov/catdir/cpso/lcc.html

biology, news, or medicine. Organizing files into a hierarchy of folders in an operating system is an example of a personal classification system.

Also, thesauri may be enumerative (flat) or have a hierarchical structure based on "broader" or "narrower" relations. In addition, the terms may have other semantic relationships, such as partitive and associative relations. Semantic relations are used by indexers for finding appropriate indexing terms, and by information searchers for formulating their queries accordingly. This idea is different from classifications, where the primary use case of the hierarchy is to use it for grouping documents. As a result, hierarchies in thesauri are often only partial, while classifications are monolithic and complete (i.e., every category has at least one supercategory except the most general root category). Another difference between classifications and thesauri is that while thesauri terms typically refer to atomic concepts, such as "art," the meaning of classification categories can be more complicated and compound, such as "News and Media" in the Yahoo! directory[16]. A classification category characterizes its document set, not the category as a concept. In a thesaurus, the term with its semantic relationships can be seen as a kind of definition of the underlying concept in the context of the other terms. This idea is particularly explicit in terminology research [143].

The idea of domain ontologies adds still new flavors to the ideas of classifying things (in classifications) and describing terms (in thesauri). The purpose of ontologies is to model a domain by defining its concepts and their relationships in a way that is agreed upon within a community and has explicitly defined syntax and semantics, so that computers can interpret the structures and infer new knowledge based on them.

Table 5.5 summarizes the differences between classifications, thesauri, and ontologies. It should be noted, however, that many knowledge organization systems may bear features from these three types at the same time.

Table 5.5: Comparing prototypical classifications, thesauri, and (domain) ontologies			
Feature	**Classification Scheme**	**Thesaurus**	**Domain Ontology**
Purpose	Organizing documents	Describing content	Defining domain concepts
Formalism	Natural language expr.	Natural language terms	Formal language
Nodes	Complex & atomic	Atomic concepts	Atomic concepts
Arcs	Document set inclusion	Standard arc types	Properties in RDF
Instances	Documents	Terms	Instances of classes
Interpretation	Humans (mostly)	Humans (mostly)	Machines and humans
Examples	DDC, LCSH, UDC	ANSI Standard thesauri	SUMO[17], OpenCyc[18]

Many thesauri and ontologies are used for defining *universals*, i.e., general classes of individuals, such as "chair" (artifact type), "wood" (material type), "painter" (actor type), or "city" (geographical feature type). Another complementary type of KOS are instance-rich ontologies or registries of individuals. Such ontologies include, for example, geo-ontologies, such as TGN, and actor ontologies (persons and organizations), such as ULAN. These kinds of ontologies of individuals are based on

---

[16]http://dir.yahoo.com/Arts/

a small ontology of classes (universals), such as "city" or "person," that is *populated* with individuals from, e.g., a database.

5.3.2 ONTOLOGY TYPES BY MAJOR DOMAINS

Major ontology types needed in CH applications can be classified by their domain of discourse as follows.

1. **General concept ontologies** These ontologies contain general concepts, such as object types (chair, painting, book, etc.) or materials (steel, wool, wood, etc.). Concepts in keyword thesauri typically fall in this category, excluding free keywords, such as place and person names.

2. **Actor ontologies** These ontologies contain lists of individual persons, organizations, and groups. In libraries, actor ontologies are called authority files.

3. **Place ontologies** These ontologies contain lists of individual places. In land surveying, place ontologies are called gazetteers.

4. **Time and period ontologies** Time ontologies specify the way in which time is represented, and may list particular periods of time for shared reference, such as "18th century," "Bronze Age," "Cambrian Period," etc.

5. **Event ontologies** Events are the semantic glue that associates actors, objects, places, and time together. Event ontologies are repositories for listing references to individuals events, such as "Coronation of Napoleon" or "World War I," so that they can be referred to in different metadata records for interoperability.

6. **Domain nomenclatures or terminologies** Different domain areas use specific nomenclatures, that roughly correspond to free keywords of thesauri. For example, there are name lists and taxonomies for plants and animals, minerals, chemical compounds, diseases, medicines, trademarks, etc.

Ontologies in all categories have foundational aspects in addition to domain specific contents. The foundational aspects cover principles of modeling that are to a large extent generic and domain independent, such as the meaning of subclass hierarchies or theories of time or parts and wholes. There are specific *foundational ontologies* focusing on these issues, such as DOLCE[19] [44]. By sharing foundational modeling principles, ontologies can be made semantically interoperable in a deep sense. Semantic Web ontology vocabularies such as RDFS, SKOS, and OWL, can be seen as simple foundational ontologies, through which shared semantics such as subclass relationship, property inheritance along hierarchies, and the meaning of basic relation types has been agreed upon.

[19]http://www.loa.istc.cnr.it/DOLCE.html

In many cases, an ontology may contain concepts from different domain categories, but the distinction above is useful for identifying concepts that are different in nature, are often maintained in different ways by different kinds of organizations, and may require different technical means for representing and using them. An ontological challenge of the CH domain is that it encompasses all domains of life. Therefore, all kinds of ontologies are potentially needed, although the needs in different memory organizations may be more specific and different, say in museums of natural history, in cultural history museums, or in art galleries.

In the following, domain ontology types and their differences are discussed using the domain-based classification presented.

5.4   ACTOR ONTOLOGIES

Actor ontologies, or authority files, serve two basic functions: First, they are used to disambiguate items with similar names. Second, they collate information related to the items. In this section, challenges of traditional authority control are first discussed and contrasted with the ideas of the Semantic Web, based on [88]. After this, pointers to a few actor ontologies are given.

Authority control includes the processes of maintaining author, title, and subject headings for bibliographic material (person, group, or organization) in a library catalog [144, 145]. The basic problems addressed here are the following.

- How to encode names referring to the same entity in a systematic way, so that the resources can later be found by searching (e.g., names can be transliterated differently in different languages)?

- How to guarantee that different entities are not encoded by a similar name, which would lead to confusion in information retrieval (e.g., different instances of "John Smith" encoded with the same string)?

Current authority file methodology relies on rigid syntax and rules for presenting the information. The basic sets of rules define the forms of the names (e.g., use the form *Surname, First Name* for persons' names), rules for changing the records when the names change, and so on. The main goal is to enable efficient search and retrieval based on names, that are often ambiguous, encoded in varying ways, and subject to change.

Traditional authority control works well on small homogeneous environments and datasets, but as a downside, it requires much expertise from the indexers and manual work. These problems are emphasized when managing and interlinking large databases, such as authority records of different libraries in different countries based on different languages. Such tasks are becoming more and more popular on the Web. At the same time, less and less experienced people are indexing content, e.g., at various Web 2.0 sites. To cope with these trends, new kinds of approaches and tools for authority control are needed.

On the Semantic Web actors are identified using HTTP URIs whose uniqueness can be guaranteed by the Domain Name System (DNS) of the WWW. The various linguistic representations of an identity are represented as literal properties attached to the URI. By transforming authority data into RDF-based format the content can be linked with related data, enriched with reasoning, and be queried using standard protocols such as SPARQL. Authority records are used in two ways. First, they are used in the traditional way for authority control, e.g., for finding unique identifiers for identities during indexing. Second, the authority RDF files can be exploited as a reusable *content repository* for applications, such as semantic portals for cultural heritage.

Authority control in libraries has traditionally had two main objectives [144].

1. Find a work (e.g., a book or an article) whose author, title, or subject is known.

2. List all works by a given author (or by subject or another attribute like genre).

These objectives are called *finding* and *collocating* objectives, respectively. More recently, the user's goals have been emphasized in the Functional Requirements for Bibliographic Records (FRBR) framework discussed in Section 4.4.5.

A typical solution to meet the requirements is to build an authorized record for each document and actor. The record contains titles (and possibly their sources) and cross references. An example of an authority record is shown in Table 5.6, taken from a requirements document by the Functional Requirements and Numbering of Authority Records (FRANAR)[20] working group.

Table 5.6: An authority record according to FRANAR
Authorized heading
Mertz, Barbara
Information note/see also references:
Barbara Mertz also writes under the pseudonyms Barbara
Michaels and Elizabeth Peters.
For works written under those pseudonyms, search also under:
>> Michaels, Barbara, 1927-
>> Peters, Elizabeth
See also reference tracings:
<< Michaels, Barbara, 1927-
<< Peters, Elizabeth

The record is identified by the authorized heading. The format of the heading is strictly defined as it glues the authorized record with the actor's actual works, such as books or articles. The additional information on the record helps the user to track related records and sources. It can also be used to disambiguate authors with similar or identical names. Automatic tools for creating authority records include clustering [41] and other name matching algorithms such as [19, 43], but even with these methods, human interaction is often required.

Authority records are often praised for their high quality. Maintaining costs are traditionally regarded as the most weighting drawback. Emerging problems include reusing records between different libraries, museums, and achieves for aggregating contents. Many of these problems originate

[20]http://archive.ifla.org/VII/d4/wg-franar.htm

from the archaic syntax used for representing the records (e.g., the MARC formats[21]), and lack of common, shared vocabularies and repositories for authority records. One of the major problems is that an authority record is a list of literal values, where the record is identified by a selected special name. Selecting the name and its form in a systematic way in a global, distributed, multilingual, temporal environment is often a tricky problem. Furthermore, there is the problem of matching the selected authority name(s) with the actual name(s) used in the library databases, where different conventions may proliferate even within a single collection by different catalogers. In many cases, authority files are intended for human usage, and are difficult to interpret by machines—-a central task on the Semantic Web.

In summary, problems of traditional authority records include the following.

1. Maintaining authority records is costly, since it requires lots of expertise and handwork.

2. Aggregating content is difficult, since different naming conventions are in use in different authority files and library databases.

3. Records evolve in time, e.g., a person may change her name, which leads to different annotations in different times.

4. Records may use complicated syntax and metadata formats that make it difficult to make contents mutually interoperable.

5. Records based on literal expressions do not link uniquely or straightforwardly with Web resources.

6. Efforts to build records are difficult to share at least on the level of the WWW.

A recent approach to overcome some of the problems with authority files is the Virtual Authority Files[22] (VIAF). It attempts to aggregate the authority files of, among others, the Library of Congress, the Deutsche Nationalbibliothek, and the Bibliothéque Nationale de France under one service. The work is controlled by a central authority and is based on MARC.

In Linked Data, various ontologies of actors are in use. A commonly used metadata vocabulary for representing persons and relations between them on the Web is FOAF[23] (Friend of a Friend). BIO[24] and Biography Light Ontology[25] are vocabularies for representing biographical information, including events. In the LOD cloud, populated ontologies extracted from Wikipedia/DBpedia are used, such as YAGO[26] [58]. Its current version, YAGO2, contains facts about some 10 million entities with 450 million facts with a manually confirmed accuracy of 95% regarding the extracted facts. The

---

[21]http://www.loc.gov/marc/
[22]http://www.oclc.org/research/projects/viaf/
[23]http://www.foaf-project.org/
[24]http://vocab.org/bio/0.1/.html
[25]http://metadata.berkeley.edu/BiographyLightOntology.pdf
[26]http://www.mpi-inf.mpg.de/yago-naga/yago/

knowledge base includes lots of actors and is linked with resources in DBpedia, GeoNames, and WordNet.

ULAN is a semantically rich and encompassing ontology of c. 120,000 artists, including some groups and organizations, too. It has a versatile system for representing alternative names of actors. Lots of useful information about persons is available, such as their professional roles (painter, architect, etc.), nationality, and a rich social network connecting persons using relations such as *child of*, *partner of*, and *student of*. ULAN is consistent with TGN and AAT, two other major vocabularies of the Getty Research Institute. RDF versions of ULAN have been created and used in several semantic CH portals, but ULAN is not available as Linked Open Data.

Recently, various libraries have published authority files (and collection metadata) as Linked Open Data, e.g., in Sweden, the U.K., Germany, the Netherlands, and Finland. In many countries, legislation and privacy concerns may prevent publication of actor data, especially regarding data about contemporary persons. A challenge of using linked authority datasets is that different metadata models are used in different publications, and that the datasets may not be strongly linked with other datasets.

5.5 PLACE ONTOLOGIES

Geospatial place ontologies define classes and individuals for representing geographic regions, their properties, and mutual relationships [87, 155]. Interoperability in terms of geographical places can be fostered by sharing place resources in different collections and application domains.

A geo-ontology can be created based on four different traditions of geography [55, 119].

1. The spatial tradition has emphasized the use of quantitative methods in geographic research; places are studied in terms of their spatial attributes, namely location, position, and geometry.

2. In the area studies tradition (regional geography) researchers have divided the world into smaller units based on their dominant features or characteristics. These units (i.e., region or area) are often subdivided into four static typologies, namely 1) formal (uniform), 2) functional (nodal), 3) administrative, and 4) perceptual region.

3. The man-land tradition entails a focus upon the relationships and interactions between societies and natural environments, i.e., how people and activities are affected and controlled by the physical environment (and also vice versa).

4. The earth science tradition lays the focus fully on the physical environment around us: the waters of the earth, landforms, vegetation, soils, topography, etc.

Geographic datasets are usually stored in *Geographic Information Systems* (GIS) facilitating efficient information retrieval based on, e.g., coordinates and maps.

There are lots of databases and repositories available for contemporary places provided by national land survey organizations and international consortia. In many cases, such data is available with open licensing. A particularly important place ontology in Linked Data is GeoNames[27]. It belongs to the LOD cloud and has been linked with various datasets in it, such as DBpedia. The GeoNames gazetteer covers all countries and contains over eight million placenames. GeoNames data is harvested from tens of other databases and there is also an interactive Web 2.0 interface for the developer community for adding places. The dataset is available for download free of charge.

Dealing with historical geographical content adds a temporal dimension and notion of change to geographic information systems. For example, a reference to "Germany" or "the U.S." may refer to different regions, depending on the time of reference. For example, Germany in 1943, 1968, and today covers quite different areas, and the question of what Germany actually was in the 18th and 19th century may be vague.

There are also gazetteers and ontologies describing historical places [136], such as Gazetteer for Scotland[28], the thesaurus of the Alexandria Digital Library[29], and the Thesaurus of Geographical Names (TGN)[30]. TGN includes 1,106,000 names and other information about places. Geo-ontologies typically include geographical feature types, such as "country," "village," "mountain," etc., individual places as their instances, a part-of hierarchy of places, a coordinate point of the place or its polygonal area, various metadata for human users, and an identifier for referencing the concept. For example, the entry for the city of New York in TGN list its various names, such as "New Amsterdam" and "Big Apple," tells its hierarchic position in the U.S. and additional larger regions (e.g., that it belongs to the state of New York), place types (e.g., city, port, national capital in 1778, etc.), and references to literal and other sources explaining, e.g., the alternative names, such as "New Amsterdam" (historical place) in more detail.

If content is annotated with a current or a historical place name and queried with the same name, stored content can be found. However, names have multiple meanings (e.g., Paris in France vs. Paris in Texas) and places can be annotated and referred to using geographically overlapping concepts with different names. In a time perspective, a region R can be referred to, in principle, by any region name at different granularity levels that has at some point of time overlapped R. For example, Helsinki in Finland can be referred to by any regional boundaries of the city since its establishment in 1550, by the various incarnations of the neighboring regions annexed to Helsinki, by different regions of Sweden before the Napoleonic wars, by Russian regions in the 19th century, by regions of independent Finland since 1917, and by EU nomenclature since 1995. A simple approach used, e.g., in TGN is to associate names with alternative names, but this is problematic when the same area or its part can be referred to by *different* overlapping places. A part-of hierarchy eases the pain w.r.t. regions and subregions, but even then there is the problem that the hierarchy is time dependent. For example, New Amsterdam has been part of the Netherlands, but is used as an alternative name for contemporary New York in TGN. The city was renamed "New York" only in 1664 by the Duke of

---

[27]http://geonames.org/
[28]http://www.scottish-places.info/
[29]http://www.alexandria.ucsb.edu/
[30]http://www.getty.edu/research/conducting_research/vocabularies/tgn/

York under the British rule. Also many other relations of regions change in time. For example, New York used to be the capital of the U.S. for a while but not anymore.

For a more accurate and machine interpretable representation of historical places, the notion of a spatio-temporal named region during a period of time is needed. Relating such regions or places ontologically with each other is useful in information retrieval, because the end-user may not use the same place names in search queries that are used in annotations, but only related place names. More generally, ontological, topological, and other relations between historical places are needed in order to link semantically related content with each other in applications.

From the perspective of the Semantic Web, this need creates challenges, such as the following.

- **Spatio-temporal ontology models**. How to represent geo-ontologies of spatio-temporal places that change in time?

- **Spatio-temporal ontology maintenance**. How to maintain spatio-temporal ontologies that change in time?

- **Annotation support**. How to support content creation using such ontologies, so that correct references to places in time can be made?

- **Application**. How to utilize such spatio-temporal ontologies in applications for querying, recommending, content aggregation, and visualization?

To address these challenges, spatio-temporal regions and their collections can be used as annotation concepts with persistent URIs, and be defined and related to each other by a time series of ontologies [73]. Regions of different kinds can be characterized from a spatio-temporal point of view by their time span and area and be related with other by topological relations [26], such as

1. the part-of relation defining hierarchies,

2. overlap relation telling how much regions overlap, and

3. other relations, such as neighbor-of, near-by, etc.

These relations are potentially useful in query expansion [16, 75] and in semantic linking on a spatial dimension. For example, when searching for castles in Europe, it makes sense to return castles in different countries that are part of Europe. However, from an IR query expansion point of view, it is not always clear when the relations can be used. For example, when querying documents about the EU, one probably is not so interested in documents about the member states but documents about the EU as a whole. Here recall is enhanced but at the cost of precision.

5.6   TIME ONTOLOGIES

Time is a central concept in the CH domain. Time references can be either linear or cyclic.

5.6.1   LINEAR TIME

Linear time references are made to

- *points of time*, e.g., the year 1492 Columbus found America or the day 1879-12-18 Joseph Stalin was born,

- *time intervals*, e.g., that World War I took place from July 28th in 1914 until November 11th in 1918, or

- *uncertain or fuzzy periods of time*, such as the Renaissance in Europe that spanned roughly from the 14th to the 17th century.

Linear time expressions, such as intervals, can be built by filling up dates and times in a metadata schema. In addition, specific concepts can be created for lexicalized time periods, such as "Bronze Age" or "Renaissance" above, and organized into ontological hierarchies with additional relations to other periods. Such structures can be found in many thesauri, e.g., in AAT under the category "Styles and Periods." A difficulty in using such expressions is that they can be culture-dependent. For example, the Bronze Age the northern Europe is quite different from that in the Mediterranean area.

Dublin Core Metadata Terms includes a time period encoding scheme[31], based on the widely used ISO 8601 standard for representing dates and times[32] summarized in Table 5.7. Here $YYYY$ is a four-digit year, MM a two-digit month, DD a two-digit day of month, hh two digits of hour (00 through 23, am/pm is not allowed), mm two digits of minutes, ss two digits of seconds, s is one or more digits representing a decimal fraction of a second, and TZD is the time zone designator (Z or +hh:mm or -hh:mm).

Table 5.7: Date and time formats according to the ISO 8601 standard

Time	Format	Example
Year	YYYY	2012
Year and month	YYYY-MM	2012-12
Date	YYYY-MM-DD	2012-12-15
Date, hours, and min.	YYYY-MM-DDThh:mmTZD	2012-12-15T19:20+01:00
Complete date	YYYY-MM-DDThh:mm:ss.sTZD	2012-12-15T19:20:30.45+01:00

A special challenge in the CH domain is inexactness of time expressions. Historical date descriptions are often incomplete, our knowledge about them is uncertain, or the dates are fundamentally vague (fuzzy) in nature. For example, if a museum catalog tells that an artifact was created in 1825–1830, does it mean that the cataloger does not know exactly the year when the artifact was constructed, was the artifact being built with varying intensity during the whole interval of time, or was the particular artifact model manufactured during the time interval? In archeology, dating findings and interpreting time expressions are particularly vague and challenging [82].

[31]http://dublincore.org/documents/dcmi-period/#sec2
[32]http://www.w3.org/TR/NOTE-datetime

An approach to representing linear time is the Time Ontology in OWL[33]. Modeling time in music is considered in the Timeline Ontology[34]. The CIDOC CRM model includes a versatile model for representing time points, intervals, and fuzzy date expressions[35].

5.6.2   CYCLIC TIME

Cyclic time references include periodical parts of a day (morning, noon, etc.), days of a week, yearly seasons (e.g., spring, summer), and expressions of opening hours. Concepts for cyclic time concepts can be found in thesauri and subject heading lists. A model of representing opening hours used in Linked Data is included in the GoodRelations vocabulary[36] (class `gr:OpeningHoursSpecification`), a lightweight ontology for annotating offerings and other aspects of e-commerce on the Web.

5.7   EVENT ONTOLOGIES

An event ontology is useful for the following types of use cases [71, 91].

- **A metadata schema**. The ontology defines a metadata model that can be used to represent historical events in CH applications.

- **A domain ontology/gazetteer**. The individual events, conforming to the metadata schema, can be used as a gazetteer for indexing historical cultural heritage content, such as a photograph taken at a particular event.

- **A data repository**. The ontology with its content can be used as a data source of its own about history, and be linked with related datasets. Additional data and descriptions about the events can be linked to the event URIs.

Several semantic models have been proposed for representing events and the relationships between them, such as the Event Ontology [118], LODE ontology [133], and SEM [151]. The CIDOC CRM is based on events, associating at its top level physical things, conceptual objects, actors, places, and time spans. Event-Model-F [130] is a foundational OWL ontology for modeling events based on DOLCE. An ontology for narrative event structures was presented in [77] and events have been used in various parts and ways in semantic CH portals [124].

There is also research on developing history ontologies and markup languages, and visualizing events on timelines and maps. A model of how history can be represented in an ontology, with map visualizations, is presented in [109], and an ontology of historical events in [70]. The Papyrus project developed ontologies for bridging the gap between CH collections and their historical attributes in news archives, based on CIDOC CRM [121]. Visualization using historical timelines is discussed, e.g., in [76], and knowledge representation of narratives in [146, 159].

[33]http://www.w3.org/TR/owl-time/
[34]http://motools.sourceforge.net/timeline/timeline.html
[35]http://www.cidoc-crm.org/docs/How_to%20implement%20CRM_Time_in%20RDF.pdf
[36]http://www.heppnetz.de/projects/goodrelations/

5.8   NOMENCLATURES

Nomenclatures are terminologies that name things within particular domain areas of sciences and other fields of life. For example, trademark databases often needed in museums fall in this category of KOS.

Biology is a science particularly rich in nomenclatures. Here systematic taxonomies for organisms have been created since the time of Carl Linnaeus (1707–1778), the father of Taxonomy. Biological *taxonomies* are hierarchical structures of *taxa* of different generality, such as Domain, Kingdom, Phylum, Class, Order, Family, Genus, and Species. It has been estimated that some 30 million species of plants, animals, and micro-organisms are living on the Earth. Taxonomies for many organism groups are freely available in RDF form, such as the multilingual Birds of the World ontology AVIO (some 10,000 taxons) and the Mammal ontology MAMO (some 5,000 taxons)[37].

Standardized name lists and taxonomies facilitate communication and are the backbone of organizing collections and observation databases in natural history museums. Taxonomies are used in the management of global biodiversity[38], too. Changes of ontological structures in time are a concern also in Biology: taxonomies change in time due to new scientific findings, opinions of authorities, and changes in our conception about life forms. Organism names and their meaning change in time, different authorities use different scientific names for the same taxon in different times, and various vernacular names are in use in different languages. This makes data integration and information retrieval difficult without detailed biological information. [147]

Nomenclatures are developed and used in virtually all sciences, such as Chemistry, Medicine, Astronomy, and Geology. Medicine has been a particularly active area in developing ontology-based nomenclatures, such as the Snomed CT (Systematized Nomenclature of Medicine—Clinical Terms) ontology[39]. This massive nomenclature contains over 311,000 uniquely identified concepts in acyclic taxonomic hierarchies with some 11,360,000 mutual relationships.

5.9   BIBLIOGRAPHICAL AND HISTORICAL NOTES

Knowledge organization systems have been developed since the times of Aristotle under different names. Different disciplines have focused on different aspects of such systems but in all areas the general trend has been toward more formal explicit models. Semantic Web represents the formal endpoint in the spectrum of approaches to KOSs with a rich set of methods and technologies available; it has the potential of becoming a harmonizing approach between different disciplines. However, there is a danger of going too far in formalizing the world; symbol structures and logic have their limits and are not a panacea. In Linked Data, a modest practical approach of using lightweight ontologies has been adopted with the idea that the model does not have to be perfect or detailed in order to be useful. It is suffient if the new approach is more useful than existing legacy systems.

---

[37]http://www.seco.tkk.fi/ontologies/biology/
[38]http://www.gbif.org/
[39]http://www.ihtsdo.org/

International standardization work thus far has focused on domain independent ontology languages and schema frameworks, such RDF(S), SKOS, OWL, CIDOC CRM, and FRBR. The number of concepts in such vocabulary systems is counted in tens or hundreds at most. In contrast, the number of concepts in domain ontologies for general concepts, places, actors, events, and names in nomenclatures can be in the millions. Aligning such systems with each other, including problems of multilinguality and data fusion for multiple copies of similar information, will be among the major challenges in the international Linked Data initiative.

Today, lots of text and handbooks about the Semantic Web ontologies are available, such as [2, 4, 34, 57, 139]. A list of them is maintained on the Web[40].

To foster the publication and use of ontologies, ontology library services have been created [112], such as Bioportal [111], ONKI[41] [154], and Cupboard [30]. Ontology Library Servers can be used in cataloging systems for finding and fetching concepts, e.g., for creating mash-up annotation applications. Such services can be used in information retrieval, too, and in supporting ontology development processes.

---

[40]http://www.w3.org/wiki/SwBooks
[41]http://www.w3.org/2001/sw/wiki/Books

CHAPTER 6

Logic Rules for Cultural Heritage

Logic and rule systems are specified in the Semantic Web layer cake model on the next layer above ontologies. However, the semantics of the metadata and ontology layers are also based on logic. The key idea of logic is to make it possible to derive new knowledge and solve problems based on known facts by mechanical inferencing. In this sense, logic is similar to arithmetics that is used for similar purposes in numerical problem solving.

This chapter first presents the classical idea of syllogisms as a background for modern logic. After this, logic used on the Semantic Web are introduced. In conclusion, the potential of utilizing logic in processing CH content in Web portals is considered.

6.1 THE IDEA OF LOGIC

Aristotle can arguably be considered the first ontologist. In addition, he was arguably also the first logician, due to presenting the first mechanical reasoning system. This system is based on four types of *propositions* and *syllogisms*, i.e., inferencing rules based on propositions. Propositions are generic patterns for stating facts: Universal Affirmative (A) "All A are B," Particular Affirmative (I) "Some A is B," Universal Negative (E) "No A is B," and Particular Negative (O) "Some A is not B."

Syllogisms are reasoning patterns of form

$$C_1, \ldots, C_n \rightarrow H$$

that can be interpreted as: if condition propositions (premises)[1] $C_1, \ldots, C_n$ hold, then the conclusion proposition H holds, too.

Some syllogism patterns capture human reasoning logic. For example, using the pattern "A, $A \rightarrow A$"

$$\text{All A are B, All C are A} \rightarrow \text{All C are B}$$

the following result about the mortality of Greeks can be reasoned with A="humans," B="mortal," and C="Greeks":

$$\text{"All humans are mortal," "All Greeks are humans"} \rightarrow \text{"All Greeks are mortal"}$$

[1] Aristotle used only two premises.

The same idea is in use in Linked Data using modern logics: metadata and ontologies are described with propositions, and rules are used for enriching the database with new facts.

6.2   LOGICAL INTERPRETATION OF RDF(S) AND OWL

The meaning of basic Semantic Web languages RDF, RDF Schema, and OWL is defined in terms of classical first-order *Predicate Logic*. This simple logic forms the foundation of logic systems, and restricted forms of it can be processed efficiently by various algorithms of computational logic. Semantic Web languages are defined with a logical interpretation that tells precisely what the underlying RDF graphs and constructs mean and how they can be used for inferencing.

The basic logical interpretation of an RDF graph is simple: each triple <*subject, predicate, object*> (e.g., <John, parent, Mary>) can be seen as a proposition fact $predicate(subject, object)$ (e.g., parent(John, Mary)), and an RDF graph is a set of facts, a knowledge base. W3C has published a formal specification[2] of a precise semantics for RDF and RDFS, and a corresponding complete system of inference rules. The semantics of RDF(S) constructs, such as `rdfs:subClassOf`, can be defined by *entailment rules*. They are of the same form that was used above for representing syllogisms: If an RDF knowledge base E contains triples $C_1, \ldots, C_n$, then triple(s) H can be added into E as new known facts. Two RDFS entailment rules specifying the meaning of class hierarchies are listed in Table 6.1 to illustrate the idea.

	Table 6.1: Two entailment rules for RDF(S)	
Rule	**If knowledge base E contains**	**then add:**
rdfs9	x `rdf:type` y, y `rdfs:subClassOf` z	x `rdfs:subClassOf` z
rdfs11	x `rdfs:subClassOf` y, y `rdfs:subClassOf` z	x `rdfs:subClassOf` z

These rules mean, for example, that if the knowledge base contains the facts

```
// Instance data :chair1324 rdf:type :backstool .
```

```
// RDFS Ontology :backstool rdfs:subClassOf :chair .
:chair rdfs:subClassOf :seating_furniture .
:seating_furniture rdfs:subClassOf :furniture .
```

then one can add the following new triples into the knowledge base:

```
:chair1324 rdf:type :chair . :chair1324 rdf:type :seating_furniture .
:chair1324 rdf:type :furniture .
```

```
:backstool rdfs:subClassOf :seating_furniture .
:backstool rdfs:subClassOf :furniture .
:chair rdfs:subClassOf :furniture .
```

The inferred new knowledge, e.g., that backstools are pieces of furniture, can then be utilized when searching for, e.g., "furniture."

[2]`http://www.w3.org/TR/2004/REC-rdf-mt-20040210/`

Reasoning can be performed either dynamically during searching, or in advance in a preprocessing and data indexing phase. Dynamic reasoning is challenging from a computational efficiency perspective when dealing with large datasets. Preprocessing means that the indexes have to be updated after changes in the knowledge base, and that the triple store grows bigger. However, current search engines, such as Apache Lucene/Solr[3], scale up efficiently to deal with very large datasets of even billions of triples.

The OWL ontology language with its various dialects provides the end-user with more expressive logical constructs for specifying ontologies. This means that more inferencing can be performed than by using more light-weight languages such as RDF(S). The logical foundations of different OWL dialects [57] are based on Description Logics and are beyond the scope of this book. OWL constructs can be used by an ontologist in a declarative way without a detailed understanding of how the underlying OWL reasoner actually works. In Linked Data usually only simple ontological models are used in practice.

A typical use scenario of developing an OWL ontology is the following. First, the ontologist develops class definitions based on a central subclass hierarchy with minimal multiple inheritance. After this, an OWL reasoner is run, enriching the ontology with additional triples and structure, such as additional hierarchies. The ontologist then checks the enriched ontology, whether the new structures are indeed correct, and edits the original ontology if needed until the result is satisfactory. The enriched ontology can be used in applications.

6.3   RULES FOR REASONING

Ontology modeling and reasoning in RDF(S) and OWL focuses on inferring terminological or ontological relations, such as instance-class relationships (rdf:type), subclass relationships (rdfs:subClassOf), subproperties (rdfs:subPropertyOf), and equalities. However, CH collections involve not only such generic relations but also domain specific relations of virtually any kind. Generic ontological reasoning can be extended with domain or application specific logical rule systems.

The most important logical system for specifying domain specific rule systems is Horn Logic (HL). HL is a restricted form of first-order Predicate Logic dealing with only *Horn clauses*. A Horn clause is a clause that contains at most one positive literal.[4] A *literal* in logic is either an *atomic formula*, i.e., a formula such as *parent(John,X)* that does not have a deeper propositional structure based on connectives (negation, conjunction, etc.). Literals in logic should not be confused with the notion of literals in RDF. An important property of Horn clauses is that they can be represented using the rule format $C_1, \ldots, C_n \rightarrow H$ introduced above that is interpreted as a logical implication. Then the conditions C_i and the conclusion H called the *head* are positive literals. For instance, in the example below, the predicate relations *brother*, *sister*, *uncle*, and *grandmother* are defined in terms of other predicates used as conditions.

---

[3]http://lucene.apache.org/
[4]A *clause* in logic is a disjunction of literals.

```
male(X), parent(P,X), parent(P,Y), notSame(X,Y) → brother(X,Y)
female(X), parent(P,X), parent(P,Y), notSame(X,Y) → sister(X,Y)
brother(X,P), parent(P,Y) → uncle(X,Y)
mother(X,P), parent(P,Y) → grandmother(X,Y)
```

If a Horn clause has no conditions, then H is considered a *fact*. In the family relation example above, facts could be used to represent an actual family tree:

```
→ male(John)
→ male(Bill)
→ female(Mary)
→ female(Jane)
→ parent(John,Mary)
→ parent(John,Bill)
. . .
```

If there is no head H in a Horn clause, then the conditions form a *goal*.

A goal can be interpreted as a *query* whose answers are value substitutions to the variables used in the query in a similar way as graph patterns are used in SPARQL queries. For example, the query *brother(X,Y)* → in the example above would return all brother pairs (X,Y) based on the facts and the rules.

There are efficient proof methods for solving HL queries. HL is actually used as the basis of a specific programming paradigm, *Logic Programming*, whose most notable programming language is *Prolog*. A logic program consists of a set of HL clauses and a query.

Logic Programming is a natural choice for representing RDF data and reasoning with it: facts *predicate(subject, object)* correspond to triples *<subject, predicate, object >* as in the logical specification of RDF. By using HL rules, reasoning based on the facts and other rules can be performed.

HL and Logic Programming would make a nice basis for the rule layer in the Semantic Web layer cake model. However, HL is not fully compatible with OWL dialects. They are based on *Descriptions Logics* that are other kind of subsets of Predicate Logic. In addition, some fundamental assumptions made in Logic Programming and Description Logics are different. Below, these issues are briefly discussed.

6.3.1   HORN LOGIC VS. DESCRIPTION LOGICS

Description Logics, underlying OWL, and Horn Logic, underlying rule systems, are subsets of Predicate Logic. A key question here is, what is their relationship? If, for example, OWL could be represented in HL, then the legacy of Logic Programming could be reused in Semantic Web programming. On the other hand, if HL could be represented in OWL, then rules could be incorporated in ontologies in a natural way.

It turns out that Description Logics underlying OWL and HL overlap each other only partly, as depicted in Figure 6.1. Horn Logic Programs are the intersection of First-order Predicate Logic and Logic Programs. Description Logics are a subset of First-order Predicate Logic, and their intersection with Horn Logic Programs is called Description Logic Programs. There are propositions

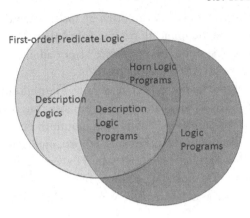

Figure 6.1: Logic systems and their overlapping [4].

that can be expressed easily in Description Logics, but not in Horn Logic. On the other hand, some propositions cannot be expressed in DL but are easy to model in HL. For example, the fact that people are exclusively either men or women can be represented in OWL easily using disjoint classes, but not in HL.

Combining DLs and rules is tricky and still somewhat under development [57]. There are two basic approaches for this.

1. Use only Horn Logic Programs. Then it is possible to incorporate full HL and rules into Semantic Web programming. Much of OWL can be represented and transformed into HL, but unfortunately not all. This choice leads to using Description Logic Programs for ontologies (cf. Figure 6.1).

2. Extend OWL with simple rules. This approach is already in use in OWL 2 where simple new predicates can be used by chaining others. For example, $grandParent(X, Y)$ can be defined based on two $parent(X, Y)$ predicates. A problem here is that although it is possible to formulate conditions on what kinds of rules can be used in OWL expression, such formulations are not necessarily simple to use from a human (or machine) perspective.

Even if HL and DL were combined successfully, interoperability of HL and DL remains a challenge, because there are two foundational assumptions made in Logic Programming that are not used in OWL: Closed World Assumption and Unique Name Assumption.

6.3.2   CLOSED WORLD ASSUMPTION

Logic Programming and Prolog usually makes the *Closed World Assumption* (CWA).

CWA means that facts that are not known to be true are assumed false by default, i.e., it is assumed that all facts about the world are known. This is a very powerful and useful assumption typically made in databases, too. CWA can often be made when dealing with CH collection data. However, if the knowledge in the application is not closed, then CWA leads to erroneous conclusions. Using CWA makes a logic system *non-monotonic*[5], since adding a new fact may falsify previously drawn default conclusions. HL itself is monotonic.

CWA is specifically not made in OWL reasoning. The philosophical reason for this is that Web knowledge in the large and by its very nature cannot be closed but is more or less incomplete. Leaving CWA out makes OWL reasoning weaker and even unintuitive in some situations, where the CWA is intuitively and automatically made by humans. For example, if we only state the fact that John is a man, we know by CWA that he is not a woman. However, using the Open World Assumption, John may still be a woman, too, unless disjointness between men and women is explicitly stated. In OWL, knowledge implicitly assumed using CWA has to be explicated by additional statements.

6.3.3   UNIQUE NAME ASSUMPTION

Another foundational difference between OWL reasoning and Logic Programming is the *Unique Name Assumption* (UNA) made in Logic Programming and database systems. UNA means that if two objects have the same identifier (name) they are equal, and conversely, objects with different names are not the same. This assumption is very natural and useful in closed systems and is a norm in databases. However, in OWL this assumption is not made. The philosophical motivation for this is that in the distributed Web context, it may well be the case that same names are used for different objects and different names for the same thing. Not using UNA can lead to unintuitive results in cases where the human implicitly (falsely) makes the UNA.

As with CWA, the OWL approach is based on pure logic without additional assumptions such as UNA. Uniqueness of names can be modeled by adding more statements in the database. For example, if a triple store contains ten identifiers for persons (:John_Smith, :Mary_Jones, etc.) and we want to tell that there really are ten different persons, one can add an `owl:differentFrom` arc between all pairs of persons. In a set of n names, this means adding $n * (n - i)/2$ arcs, in our case 45 pieces of them. In practice, this can be expressed concisely by using the `owl:allDifferent` construct instead of writing down the 45 triples explicitly. However, it is clear that using UNA, when appropriate, can simplify models a lot.

6.4   USE CASES FOR RULES IN CULTURAL HERITAGE

A collection of cultural metadata records and related ontologies constitute a knowledge base. Rules can be used for deriving new facts based on it. Some examples illustrating different ways of using rules in semantic cultural portals are given below.

- *Enriching semantic content.* Common sense rules may be used for enriching annotations, thus extending the machine's understanding about culture. For example, the family relation rules

---

[5]Reasoning in a monotonic logic, such as Predicate Logic, can only add new facts into the knowledge base, not remove old ones.

above can be used to explicate implicit family relations of persons in order to find content related to families.

It seems that many metadata records in memory organizations are partly redundant in practice. For example, an artifact may be described both as a *chair* and as a piece of *furniture*. Redundant annotations that can be produced automatically by the machine by inferencing are not needed, which makes cataloging easier for humans.

Metadata formats may contain implicit useful knowledge embedded in the relational meaning of the elements. For example, a record from an art gallery may tell that a *dc:creator* of a painting produced in Tokyo is John Lennon. One can then infer that there was a painting event in Tokyo with John Lennon as an actor, and this event could be added, e.g., into the biographical description of events in John Lennon's life.

- *Data alignment and linking.* Rules are useful when aligning and linking datasets with each other. For example, rules can be used in a natural way for identifying and merging multiple occurrences of the same resources (e.g., persons, places, or events), a recurring problem encountered when linking datasets.

- *Semantic recommendations with explanations.* A simple way to produce recommendation links in a portal is to use SPARQL queries. For example, if the end-user searched for books written by Victor Hugo, then one can recommend more books from other French authors from the same era with similar themes. Using more specific rules logics for recommending can be specified, based on traversing the underlying RDF graph. As in expert systems, chains of applied rules can be used for constructing human-readable explanations on why the recommendations are suggested.

- *Projecting search facets.* In faceted search, rules can be used for constructing facet hierarchies based on ontological structures, such as the subclass-of and part-of-relations. Furthermore, rules can be used to solve the problem of projecting search items to facet categories, which may be complicated. From a software engineering viewpoint, using logic rules for projections separates facets from the annotation ontologies and annotations, which makes it possible to apply the same faceted search engine to knowledge bases based on different kinds of ontologies and annotation schemas.

- *Association discovery.* Association discovery, also called *relational search*, can be based on rules trying to find paths between resources in an RDF graph, such as a friend-of-a-friend (FOAF) network.

When using rules, the application system does not need to be *complete* in the strict logical sense, i.e., capable of drawing all possible correct conclusions. One can be happy with a system that can infer at least some useful additional conclusions, if not all. However, application systems should be as *sound* as possible, i.e., the system should not draw wrong conclusions.

6.5   BIBLIOGRAPHICAL AND HISTORICAL NOTES

History of logic and semantic nets from a computer science perspective is discussed, e.g., in [137]. The logical foundations of the Semantic Web languages are presented in [4, 57] and originally in the W3C documentations of RDF(S)[6], OWL [7], and Rule Interchange Format (RIF)[8], a format that allows logic rules to be exchanged between rule systems.

Semantic Web systems enrich their data using at least simple ontological inferencing, such as RDF(S) entailment rules. Logic programming rules for enriching CH collection data, for facet projection, and for producing recommendations with explanations were used in [63]. Another Prolog-based CH system is [132]. Theory of Logic Programming is presented [92] and Prolog programming in practice is introduced in various textbooks, such as [22]. A free open source Prolog system with support for RDF processing, Web servers, and Semantic Web programming is SWI-Prolog[9].

Explication of knowledge embedded in metadata schema relations in terms of events was suggested in [124]. Rules are used for entity resolution and data linking in the LDIF framework [131]. Association discovery (relational search) is used, e.g., in [64, 132, 134].

---

[6]http://www.w3.org/TR/2004/REC-rdf-mt-20040210/
[7]http://www.w3.org/TR/2009/REC-owl2-rdf-based-semantics-20091027/
[8]http://www.w3.org/TR/2010/REC-rif-bld-20100622/
[9]http://www.swi-prolog.org/

CHAPTER 7

Cultural Content Creation

This chapter discusses the content creation process for a semantic CH portal, based on the model presented in Chapter 2. Here data from distributed content providers and data sources are aggregated into a global data source, data integration is performed based on an ontology infrastructure, and the content is then provided for human end-users and other Web services. Creating such a portal involves three separate but interacting content creation processes.

1. Vocabulary and ontology creation by domain communities (art, history, etc.)

2. Local content creation by the content providers (museums, libraries, archives, etc.)

3. Global content aggregation and integration by the portal publisher

   In this chapter, these processes are discussed. In conclusion, quality issues of linked data are considered.

7.1 VOCABULARY AND ONTOLOGY CREATION

The contents of a semantic CH portal are represented using a set of metadata schemas for collection items, and on a set of domain ontologies for filling in schema element values. In order to promote interoperability across distributed content creators and domains, such structures should be shared among collaborators as openly as possible. To support this, centralized services, such as ontology servers, can be utilized to provide up-to-date versions of the vocabularies to everybody in real time. Centralized services can also be used for sharing implementation work and costs between institutions.

   The users of an ontology infrastructure need to agree upon ontologies as well as policies and practices on using them, such as, cataloging guide lines.

7.1.1 CONCEPTUAL LEVELS OF ONTOLOGY CREATION

An ontology infrastructure includes ontologies at different conceptual levels.

1. **Domain independent vocabularies** are needed for facilitating cross-domain interoperability. For example, thesaurus standards and the W3C Semantic Web recommendations RDF(S), SKOS, and OWL fall into this category, as well as generic metadata schemas, such as Dublin Core.

2. **Domain specific ontologies** are created and used based on domain dependent needs and models. For example, the Getty Vocabularies (AAT, TGN, and ULAN), the Library of Congress

Subject Headings (LCSH), and other vocabularies used for annotating contents fall in this category.

3. **Institution specific ontologies** are needed for concepts that may be relevant for a particular organization only or cannot be shared for some reason with a larger community due to, e.g., privacy or copyright issues. For example, international authority files for persons may not include individuals that may have local importance in a village or town, but that are not of much international interest. Such ontologies can be created as extensions of more general vocabularies.

General domain independent vocabulary and ontology specification work is usually coordinated by international standardization bodies, such as W3C and ISO. For domain specific harmonization work, area-specific organizations have taken the lead, such as ICOM CIDOC for museums, IFLA for libraries, and ICA for archives. Their recommendations are then adapted and translated for local use by national libraries and other national organizations developing, e.g., thesauri and other knowledge organization systems for different domains. Institution specific ontologies and ontology extensions are developed by particular museums, libraries, or archives, or individual persons.

Domain ontologies can be created from scratch using ontology editors, such as Protégé[1]. Methods and tools for automatic or semiautomatic ontology construction, i.e., for *ontology learning* have been developed, too. Creating ontologies, either manually or using (semi)automated means, is an art of its own and will not be discussed in depth in this book. Pointers to literature are given in the last section of this chapter.

Ontologies used in Linked Data are usually light-weight and based on legacy thesauri and vocabularies that have been transformed more or less mechanically from existing data sources. In the next subsection, transformation of keyword thesauri into RDF is discussed to illustrate the process and semantic issues involved.

7.1.2   TRANSFORMING LEGACY THESAURI INTO ONTOLOGIES

Consider a traditional thesaurus [1, 40] that is based on the semantic relations of Table 5.2. There are two obvious possibilities for transforming it into RDF. 1) Using SKOS. 2) Using RDF(S) and OWL.

Transforming Thesauri into SKOS

When transforming a thesaurus into SKOS, terms are represented as instances of the class skos:Concept or their collections. BT/NT relations can be represented using the properties skos:broader/skos:narrower. Transforming a simple standard thesaurus into SKOS is fairly straightforward. However, CH vocabularies, such as Iconclass, WordNet, ULAN, AAT, TGN, etc., are typically more complex in terms of both structure and semantics. Analysis and transforma-

---

[1]http://protege.stanford.edu/

tion of such vocabularies is discussed in [148, 149, 150]. Various SKOS-transformations have been published on the Web and have been used in cultural semantic portals [132, 152].

Although a syntactic automatic transformation of a thesaurus into SKOS is useful, the semantic ambiguities and inaccuracies of the original thesaurus will prevail in the RDF SKOS version. A fundamental problem with many traditional thesauri is that their semantic relations have been created mainly to help the indexer in finding indexing terms and to help the information seeker in formulating queries accordingly. Interpreting the meaning of the relations in a thesaurus, such as Broader/Narrower/Related Term, requires human understanding about the underlying world. Unless the meaning of the semantic relations of a thesaurus is made more explicit and accurate for the computer to interpret, the RDF version is equally confusing to the computer as the original thesaurus, even if Semantic Web standards are used for representing it.

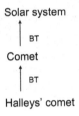

Figure 7.1: An example of a broader term hierarchy.

For example, consider the terms of Figure 7.1 taken from the YSA thesaurus[2]. One can easily understand their meaning but the machine would be confused without additional knowledge: The concept *Comet* obviously represents the class of comets, but how would a computer know whether *Halley's Comet* is an individual or a class? *Comet* is a subconcept of *Solar system* but how would a computer know whether comets are a kind of solar system or part of a solar system? And in what sense are comets part of solar systems, given the different meanings of part-of-relations listed in Table 5.3? Such semantic distinctions may be necessary when using the annotations in portal applications. For example, if the BT relation is used as it is for query expansion, then a search for "solar systems" would falsely retrieve comets, although comets are not solar systems. By refining BT relations into either subclass-of or part-of relations such problems can be avoided. This does not mean that the ambiguous BT relations are useless: when searching for documents related to "solar systems" documents related to "comets" could be acceptable in the result set. However, removing ambiguities makes the vocabulary more useful for the computer.

Defining the meaning of indexing concepts accurately is essential when enriching semantic content by reasoning, e.g., via property inheritance along `rdfs:subClassOf` hierarchies. Enriched contents are useful in many application areas, such as semantic search, linking, and recommending, and foster semantic interoperability. Benefits can often be obtained with little extra work, e.g., by

[2]General Finnish Thesaurus maintained by the National Library of Finland.

just systematically analyzing multiple meanings of thesaurus entries, and by organizing the concepts along subclass-relations into a hyponymy.

Transforming Thesauri into RDFS/OWL

The original W3C recommendations for representing ontologies are RDFS and OWL. When transforming a thesaurus into these formats, common terms, such as "table," "painting," or "spring," can be represented as RDF(S) or OWL classes, the BT/NT relations as a `rdfs:subClassOf` hierarchy, and other relations with appropriate RDF properties. Concepts referring to individuals (e.g., "World War II") can be represented as instances of the classes (e.g., "war"). The idea of using term concepts as classes of real world things is quite different from SKOS, where terms are instances of concepts and the focus is on representing thesauri constructs in a harmonized way rather than modeling the real world.

As in SKOS transformations, transforming the terms into classes and instances, and relations into properties, is a straightforward task to do for a computer on a syntactic level. However, the resulting structure will then contain the same semantic ambiguities as the original thesaurus unless human editing (or artificial intelligence) is involved. Below, a minimal list of semantic refinements based on [68] is presented for transforming a simple thesaurus into a subclass-of hierarchy.

1. **Disambiguate concept meanings**. Terms in a thesaurus can in practice be ambiguous and cannot be related properly with each other in the hierarchy using the subclass-of relation. For example, in YSA there is the indexing term "child." This term has several meanings such as "a certain period of human life" or "a family relation." For example, George W. Bush is not a child anymore in terms of age but is still a child of his mother, Barbara Bush. The computer cannot understand this and is confused, unless the meanings of "child" are separated and represented as different concepts (with different URIs) in different parts of the ontology.

   Legacy thesauri contain in practice lots of ambiguous terms that cannot be placed in one place in the hierarchy. In such cases the term has to be split into several concepts. However, a general ambiguous concept encompassing several meanings, say "child," can be useful for indexing purposes. For example, assume a painting depicting children playing in a park with their mothers watching. When selecting keywords (concepts) describing the subject, it would be tedious to the indexer to consider all the meaning variants of "child," while the single ambiguous indexing term "child" would encompass them properly in this case. Therefore, one may consider some useful ambiguous concepts, such as "child," to be used as special aggregate indexing concepts. They lay outside of the subclass-hierarchies but can be defined in terms of their concepts by using, e.g., Boolean class expressions as in OWL.

2. **Aggregate terms into a systematic subclass-of hierarchy**. The BT/NT relations do not necessarily structure the terms into a full-blown hierarchy but into a forest of separate smaller hierarchies. In the case of YSA, for example, there are thousands of terms without any broader term. Many interesting relations between terms may be missing in thesauri, especially concern-

ing general terminology, where BT relations are not commonly specified in practice. Adding such relations and organizing the terms into a full-blown systematic structure makes the vocabulary better linked and more usable for searching and visualizing contents.

When transforming a thesaurus into a class hierarchy, a central structuring principle in constructing the `rdfs:subClassOf` hierarchies is to avoid multiple inheritance across distinct upper ontology categories. For example, the class of *Museum* cannot be at the same time a point of interest and an organization, otherwise its instances would be at the same time places and organizations. An approach and methodology for structuring the upper categories is DOLCE [44], where the concepts on the top level are divided into three upper classes: Endurant (changing things, such as activities), Perdurant (stable things), and Abstract (e.g., measurement units).

It is also possible to organize the concepts into part-of hierarchies (*partonomy, meronomy*), and verb-like concepts into a *troponymy* of more generic concepts. For example, "walking" is more generic than "limbing," and "talking" more generic than "whispering" or "lisping." This kind of model is in use in, e.g., in WordNet.

3. **Disambiguate BT/NT relations**. The semantics of the BT/NT relation is ambiguous: it may mean either subclass-of relation, part-of relation (of different kinds), or instance-of relation. This hinders the usage of the structure for reasoning [47]. For example, the BT relation cannot by used for property inheritance because this requires that the machine knows that BT means subclass-of and not, e.g., part-of relation. To facilitate this, existing BT relations has to be refined into subclass-of and part-of relations, or instance-of relations if the distinction between instances and classes is needed.

Deciding what terms should be instances and what terms are classes is sometimes tricky. For example, should the material of a dress in a collection, say "wool," be an instance or a class? Obviously, different dresses are made of wool from different animals, but generating different wool instances for all artifacts made of wool clutters the database. This question is quite essential when using OWL where systematic use of instances as property values is needed for efficient OWL reasoners[3]. One possible way of avoiding commitments at the ontology level is to also represent instance-like terms in a thesaurus as classes, and let the applications to decide whether to instantiate such classes or use class references.

4. **Make subclass-of-relations transitive**. The transitivity of the BT relation chains is not guaranteed from the instance-class-relation point of view, when transforming BT relations into subclass-of relations. If x is an instance of a class A whose broader term is B, then it is not necessarily the case that x is an instance of B, although this is a basic assumption in RDFS and OWL semantics if BT is regarded as the subclass-of relation. For example, assume that x is a "make-up mirror," whose broader term is "mirror," and that its broader term is "furniture."

---

[3]Cf. `http://www.w3.org/TR/swbp-classes-as-values/`, W3C Working Group Note 5 April 2005.

When searching for the concept "furniture" one would expect that instances of furniture would be retrieved, but in this case the result would include x and other make-up mirrors, if transitivity is assumed. This means, e.g., that term expansion in querying cannot be used effectively based on the BT relation in a thesaurus. Refining the hierarchy relations is therefore needed.

When transforming a thesaurus into an ontology, it is likely that changes in the meaning of the terms will occur based on splitting the concepts and aligning them into subclass-of hierarchies. To provide full compatibility with existing datasets annotated using the original thesaurus mappings between thesaurus terms and concepts in the ontology may be needed.

Cross-domain Collaboration

Creating an ontology infrastructure for CH includes developing a variety of interoperable ontologies in different domains. Smooth collaboration is therefore needed between domain expert groups developing vocabularies in different areas. To foster collaboration the following guidelines are worth considering.

1. **Add machine semantics**. Start transforming thesauri into machine interpretable (lightweight) ontologies in order to boost their usage on the Semantic Web.

2. **Proceed in small steps**. Adding even little semantics can be very useful (and keeps, e.g., the funding agencies happy).

3. **Think cross-domain**. Consider not only your own micro world but also cross-domain usage of concepts when making ontological decisions.

4. **Establish collaboration networks**. Nobody masters the whole universe alone. The work must be based on independent domain expert groups.

5. **Reuse others' work**. This saves effort and enriches one's own work.

6. **Maintain interoperability** with the past and other ontologies; otherwise benefits of collaboration are lost.

7. **Respect different ontological views**. Encourage reuse but do not force the others to obey your own choices. It is not possible to come up with only one ontological view of the world. Therefore, each group should have the freedom of making its own choices (using the shared principles).

8. **Accept imperfect models**. The ontology may never be fully perfect but can still be useful.

9. **Minimal ontological commitment**. Keep ontological structures simple in order to facilitate their reuse in different domains and applications. Application-dependent structures can be created as add-ons in applications.

10. **Coordinate the work**. Ontology work is not trivial and high quality work can hardly be done without coordination.

7.1.3   TERMINOLOGY CREATION

Above, ontologies were considered mainly from the point of view of a machine. The terminology for human use in a portal is typically defined by associating ontological resources with preferred and alternative labels, using properties such as `rdfs:label`, `skos:prefLabel`, and `skos:altLabel`. Resource identifiers (URIs) of concepts, used by the machine, refer to concepts that are in principle language independent. However, labels used by humans can be multi-lingual, based on XML markup (e.g., `xml:lang`). The distinction between resources and their labels in different languages facilitates creating multilingual portals.

The content providers often use different literal terms to refer to the same resources when describing metadata in legacy systems. For example, literals "United States" and "US" may be used to refer to the same country. This problem of synonymy can be approached by using alternative labels in each language separately. On the other hand, the same term may be used to refer to different concepts, such as river "bank" and financial "bank." In order to eliminate such homonyms in terminology, it is advisable that an ontology uses a unique labeling of preferred terms for concepts in each language. Unique labeling does not solve the problem of disambiguating meanings of homonymous words occurring in natural language descriptions, but at least the terminology of an ontology should be made unambiguous. Label disambiguation can be made using qualifiers in brackets, e.g., "bank [organization]."

7.1.4   ONTOLOGY ALIGNMENT

Several domain ontologies are often used in describing cultural metadata even within a single domain. Then multiple identifiers (URIs) will be in use for denoting a single concept even if this is not desirable in general. For example, registries of same geographical locations are maintained at different countries and by different service providers using their own identifiers. This raises up the problem of making the ontologies mutually interoperable. There are several solution approaches for this:

1. ontology alignment, also called ontology matching and mapping,

2. ontology merging,

3. sharing foundational ontologies, and

4. sharing a top ontology.

Ontology alignment means that a mapping between concepts and properties in two ontologies is created [37, 50, 110]. In Linked Data the mapping is typically represented as a set of `owl:sameAs` arcs. Also other relations than strict equality, such as `rdfs:subClassOf`, or `skos:narrowMatch`, can be used in mappings.

In *ontology merging* such a mapping is used to create a new joint ontology based on the resources of the ontologies to be merged.

Foundational ontologies are formal representations of some aspects of reality, such as a theory of time or wholes and parts. If two ontologies are based on similar theories or principles they can be aligned with each other accordingly.

Using a shared *top ontology* means that vertical domain ontologies can be made interoperable via shared concepts in an upper ontology. This idea has been applied, for example, in the IEEE SUMO[4] ontology containing some 1,000 concepts and 4,000 assertions (including 800 rules) based on them. In the FinnONTO light-weight ontology infrastructure [68], there are 15 different domain ontologies that share a central top ontology of some 24,000 concepts. This system is illustrated in Figure 7.2, where YSO is the top ontology (transformed from the YSA thesaurus) and its classes are refined by vertical ontologies including AFO (for agri-forestry, 7,000 concepts), MAO (for museum domain, 6,800 concepts), TAO (for applied arts, 3,000 concepts), VALO (for photography, 2,000 concepts), and other domain ontologies. Transitivity of the subclass-of chains has been manually checked across ontology boundaries. The system of ontologies creates one large cross-domain ontology called KOKO of some 72,000 concepts (including duplicates in ontology intersections).

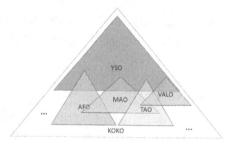

Figure 7.2: FinnONTO system of vertical ontologies sharing a horizontal top ontology.

Linked data alignment is usually based on pointing out equal and more general resources in datasets. For representing equality, the `owl:sameAs` property is typically used. Hierarchical relations between classes and properties in a dataset can be represented in a simple way by using properties `rdfs:subClassOf` and `rdfs:subPropertyOf`. The mechanism is used, for example, when developing refinements and extensions for the Dublin Core metadata model. The same approach is used also in the Vocabulary Mapping Framework (VMF)[5]. In addition, the SKOS standard includes specific properties for mapping URIs in different vocabularies onto each other.

7.1.5   ONTOLOGY EVOLUTION

Ontologies are not static but change in time. This aspect is particularly important for CH ontologies that represent concepts related to history and that evolve over long time periods. Furthermore,

---

[4]http://suo.ieee.org/
[5]http://www.doi.org/VMF/

annotations of CH contents are intended to be preserved and maintained over a long time for future generations.

Generic domain independent vocabularies are typically developed by large national or international consortia and do not change frequently. More challenging from a maintenance viewpoint are domain dependent ontologies. For example, new concepts for persons and places typically need to be created on a regular basis in museums as new data items are cataloged. After creating a new entry for, say an artist, the concept should be shared with other museums as quickly as possible. Otherwise, other museums end up creating multiple instances of the same concept and annotate their contents using them. This leads to ontology mapping problems later when aggregating content from distributed content providers.

Ontologies change due to several reasons. Firstly, changes are made to make the ontology correspond better to what it represents.

1. **Errors**. Correcting errors in an ontology leads to *ontology versioning* where later versions in the ontology series are more accurate than earlier.

2. **Accuracy**. The formal model underlying the ontology can be made more expressive (e.g., by introducing new properties) or the model more encompassing (e.g., by adding more concepts and properties).

Second, even if the ontology was correct and accurate the underlying context may change, which leads to changes in the ontology.

1. **Real world changes**. For example, places such as towns, counties, and countries change names, areas, population, etc. A way of representing such changes is to use an *ontology time series* [73, 84] where each ontology version represents the real world at a particular time or time period. Notice that this kind of time series is different from an ontology version series where each ontology version represents the same world with increasing accuracy.

2. **Human conceptualization changes**. Human ways of conceptualizing the world change. For example, the notion of "desktop publishing" was once popular in computer science, and lots of documents have been indexed with the term in libraries. Nowadays, the concept has disappeared even if the phenomenon is still there.

Another challenge of ontology evolution is how to deal with *free indexing* keywords (concepts). These concepts, e.g., persons, organisms, or trademarks, cannot be enumerated exhaustively as ontological concepts beforehand but are allowed to be used in indexing when using controlled vocabularies. This has led to serious problems regarding, e.g., authority file management of persons [144].

Free keywords are instantiated by individual catalogers in organizations, based on the rules of the vocabulary in use, not by the ontology developer team. This creates a challenge for the ontology infrastructure because the same concept may be needed by several organizations, and there is the danger of duplicating the work and creating multiple copies of the same concept with different identifiers (URIs). Updating institution specific ontologies is less challenging because the concepts are not shared with outsiders. However, mappings of such concepts with the shared domain specific vocabularies must be maintained.

Current vocabulary maintenance practices in museums and libraries may facilitate sharing of new ontological concepts as candidate suggestions within one organization. However, mechanisms for sharing concepts between organizations in real time are missing. A typical practice is to establish a committee for maintaining a thesaurus in a domain. The users of the thesaurus are urged to propose new terms or changes to the committee that evaluates them, makes appropriate updates, and publishes every now and then a new version of the thesaurus. This may be appropriate for slowly evolving domains, but not for all.

When a previously unknown free indexing concept is encountered, say a new person, the corresponding ontology needs be updated in real time, so that it can be referred to later on, when the concept is encountered again. Furthermore, the new concepts should be visible not only to their creators, but to the whole community using the ontology. So, a real-time feedback mechanism for sharing new concepts created at an institution level is needed. At the same time, a distinction between officially approved concepts in the ontology and pending concepts suggested by an individual cataloger has to be made in order to keep the process in control.

In [42] the following model for updating and sharing new concepts is proposed. The local annotation system is based on a shared ontology service. When a new free keyword is needed, the user is able to create a new entry in the ontology with a pending approval status. This means that the new concept can be used immediately by everybody using the ontology service, but everybody can also see that the new concept is not part of the official ontology version yet. When a new pending concept is created, the creator specifies a literal description of the term as well as its key relations to the host ontology, e.g., that it is a subclass or an instance of another class in the ontology. Later on, the ontology maintenance committee can check whether the new term is appropriate, possibly edit its definition, and accept it into the ontology. Annotations created during the acceptance process are based on the original persisting URI and correct even if the concept is edited and its definition completed later by the committee.

7.2   TRANSFORMING LOCAL CONTENT INTO RDF

Transformation of a local database into linked data involves two major phases: 1) transforming the local data sources of the portal into RDF form and 2) integrating the local RDF sources into globally linked data. This subsection first explains different phases of such transformations, and then discusses issues related to transforming legacy relational databases into RDF form—-this is a most common pattern of creating CH linked data in practice.

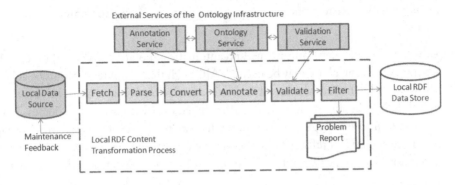

Figure 7.3: RDF content creation process for a local data source.

7.2.1 TRANSFORMATION PROCESS

The process of transforming a local legacy data source in a memory organization into a local RDF data store to be harvested by a portal is depicted in Figure 7.3. There are the following phases in the transformation.

1. **Fetch**. Data is fetched from a data source using source specific solutions. A memory organization may publish its data on a Web server as, e.g., CSV files, a database dump, in JSON, or in RDF for downloading. Alternatively, a Web service API may be used for this. In many cases, the data may be available only as HTML Web pages in which case a crawler must be used for data harvesting.

2. **Parse**. Next the data must be parsed into RDF form. Again, there are different options for doing this depending on the data source format used. There are ready to use parsers available for commonly used data formats as well as the GRDDL framework[6] (Gleaning Resource Descriptions from Dialects of Languages) for extracting RDF descriptions from Web pages.

3. **Convert**. Since the data structures used in different data sources are typically different, conversions into harmonized formats used by the portal are needed. For example, in one data source a person's name may be represented using a single literal field, while in another first and family names are separated. Also naming conventions for metadata fields must be unified, and possibly missing properties added (e.g., translations of literal labels). Unnecessary extraneous properties not to be used by the portal can be removed at this phase. After this phase the data has been harmonized on a syntactic level.

4. **Annotate**. At this phase, the property values that are usually represented as literals in legacy systems (e.g., "London," "chair") must be projected into the Linked Data space, i.e., must be

---

[6]http://www.w3.org/TR/grddl/

provided with URI references to corresponding resources in the ontology infrastructure and related datasets.

5. **Validate**. Data sources, conversions, and automatic annotation are error prone, and therefore a separate validation phase may be needed. Although the computer may not be able to automatically detect URI references based on literal description, it may at least be able to detect errors by validating the annotations against the metadata schemas and ontologies. It is often also possible to notice ambiguous situations, where a literal refers to several ontological meanings. Then the system can refrain from making the choice, and expose the problems to the human annotator. Other typical problem situations include missing data and data in the wrong format.

6. **Filter**. Results of data validation can be presented in human readable and/or machine readable form; in the latter case they can be used more easily as feedback for correcting the primary data.

7.2.2   TRANSFORMING RELATIONAL DATABASES INTO RDF

Content in memory organizations is usually available as relational legacy databases, whose annotations are literal terms and free text descriptions. In the following, transformation of such data into linked data is discussed in more detail.

The W3C RDB2RDF Working Group has specified the *"direct mapping"* method[7] for transforming relational tables into RDF format. Direct mapping is a basis for defining and comparing more intricate transformations.

To illustrate direct mapping, consider the following tables representing authorities (People, Table 7.1) and their places of birth (Places, Table 7.2).

Table 7.1: Relational table People

Primary Key			→ Places(ID)
ID	name	birthDate	birthPlace
4	Pablo Picasso	1881	18
5	Leonardo Da Vinci	1452	5

Table 7.2: Relational table Places

Primary Key		
ID	placeName	country
5	Vinci	Italy
18	Malaga	Spain

The table People lists persons, identified by a primary key, and their places of birth are represented in the table Places, using foreign keys as a reference. Given a base IRI, say `http://museum.org/db/`, the tables People and Places are represented by the RDF graph:

[7]`http://www.w3.org/TR/rdb-direct-mapping/`

```
@base <http://museum.org/db/> .

<People/ID=4> rdf:type <People> . <People/ID=4> <People#ID> 4 .
<People/ID=4> <People#name> "Pablo Picasso" .
<People/ID=4> <People#birthDate> 1881 .
<People/ID=4> <People#birthPlace> <Places/ID=18> .

<People/ID=5> rdf:type <People> . <People/ID=5> <People#ID> 5 .
<People/ID=5> <People#name> "Leonardo Da Vinci" .
<People/ID=5> <People#birthDate> 1452 .
<People/ID=5> <People#birthPlace> <Places/ID=5> .

<Places/ID=5> rdf:type <Places> . <Places/ID=5> <Places#ID> 5 .
<Places/ID=5> <Places#placeName> "Vinci" .
<Places/ID=5> <Places#country> "Italy" .

<Places/ID=18> rdf:type <Places> . <Places/ID=18> <Places#ID> 18 .
<Places/ID=18> <Places#placeName> "Malaga" .
<Places/ID=18> <Places#country> "Spain" .
```

Each row in the table is transformed into an instance of a class represented by the table (People or Places) and is identified by an IRI that is constructed by concatenating the base with the table name and a primary key value, e.g., `http://museum.org/db/People/ID=4`. Predicate IRIs are created by concatenating the base with a table name and a column name, e.g., `http://museum.org/db/People#name`. The object values are literals or IRIs used as column values in the tables.

If the database happens to have tables and columns recording the desired data, the transformation is simple. However, in practice data needed for an RDF property value in a metadata schema may be distributed in different tables and columns. For example, information about the material of an object in a museum record may be recorded in terms of keywords and a specific material field. Furthermore, the RDF model may have properties that are not explicitly represented in the museum record. For example, the technique of manufacturing an object or its artistic style may not be cataloged explicitly, but represented in terms of keywords in the *dc:subject* description.

Lots of databases have been published as linked data using this kind of transformations. However, such publications are only syntactic and not semantic, and the result is not really *linked*, if the property values are literals and not URIs, i.e., links to other pieces of RDF data. In the example above, people are linked to place IRIs, but countries are not ontological resources but only literal values. Therefore, other possible content related to, e.g., Spain, such as a vase produced there, would not be linked with Pablo Picasso, unless countries are represented as resources with IRIs, too.

When creating URI links based on literal values the following difficulties are encountered.

1. **Unknown literals and free keywords**. The values in the RDF data are likely to contain literal terms that cannot be found in the domain ontologies or associated with them. For example, free indexing terms are widely used in controlled vocabularies. Human intervention is needed for creating such concepts and for associating them with other concepts in the annotation ontologies.

2. **Spelling conventions and errors.** Another practical problem is spelling errors and morphological variants in metadata, and the variance of encoding practices used at different organizations at different times, in different languages, and even by different catalogers. For example, the name Ivan Ayvazovsky (Russian painter, 1817–1900) has 13 different labels in ULAN (Ajvazovskij, Aivazovski, Aiwasoffski, etc.), and the first, middle, and last names can be ordered and shortened in many different ways.

3. **Semantic disambiguation of homonyms.** When transforming and linking legacy metadata, a key problem is how to map ambiguous literal descriptions (homonyms) in the metadata with corresponding ontological concepts (URIs). For example, how to determine that the string "bank" in a *dc:subject* description of a photograph refers to the concept "river bank" and not "financial bank" in an ontology, assuming that such concept candidates can be found there in the first place.

When dealing with structured collection metadata, knowing the metadata element domain in which an ambiguous term has been detected can be of great help in resolving references. Interpretations of the term that do not match the element domain can be simply filtered out. For example, a river bank as a value for *dc:creator* does not make sense, but bank as an organization does. In [66] three museum collections were annotated and literal terms disambiguated in this way. Nearly 80% of homonyms could be resolved without human intervention. However, in this case ontologies for annotating the terms had already been populated based on the legacy metadata, i.e., unknown terms encountered in collection records had been recognized and inserted manually into the domain ontologies to be used in annotations, and possible typing errors were corrected.

During data transformation and linking, ontologies need to be populated with new concepts, errors in literal expressions in metadata need to be corrected, and semantic ambiguities need to be resolved. At least some human intervention is needed in these tasks. This means that the content transformation process outlined above cannot be fully automated without sacrificing data quality. Human intervention is costly, slows down the publication process, and creates the need for feeding back data corrections and refinements from the portal side to the original databases, which leads to various additional complications in the publication process. If the original data were annotated using shared ontologies these severe complications could be avoided. This is why the idea of an ontology infrastructure including services supporting annotation processes has been introduced as a central component in the portal model of Chapter 2.

The transformation process can be executed either by the local content provider or by the portal organization. In general, it is better if the local data provider, such as a museum or a library, is responsible for the RDF transformation: 1) The local data provider knows best its data and is able to process it better. 2) The data provider knows when and what should be updated in its dataset. 3) There is a feed back loop from RDF creation process to maintaining the original primary data and metadata. Ideally, the local provider could provide new versions of its dataset when needed, and request the portal organization for starting an update process concerning its data.

7.3   CONTENT AGGREGATION AND INTEGRATION

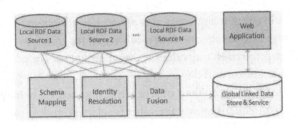

Figure 7.4: Global data aggregation of local RDF data sources.

After creating local RDF datasets, the next step is to aggregate them into a global linked data store and service. This phase is executed by the portal organization, and involves the tasks illustrated in Figure 7.4.

1. **Schema mapping** harmonizes the use of properties in the local metadata models. For example, in one dataset the subject matter may be described by a property *dc:subject* while in another dataset *:keywords* is used. After this phase properties with identical meaning are mapped onto each other, and mutual `rdfs:subPropertyOf` relations are resolved. However, schema mapping is often more difficult than just mapping individual properties onto each other. Another approach, facilitating deeper semantic interoperability of schemas, is to transform all metadata into a harmonized model such as CIDOC CRM.

2. **Identity resolution**. After harmonizing property structure there is the problem of harmonizing references to annotation ontologies used in the datasets. For example, different authority files are used in different memory organizations. This process is called *identity resolution*. The result of this phase is typically a set of `owl:sameAs` mappings of URIs in different datasets. Such mappings can be created by hand or automatically using systems that compare, measure, and identify similarities between resources based on their properties, and create corresponding mapping links. For example, two person instances having the same last name and date of birth are likely to refer to the same person, and a mapping between their different URIs can be established. If 100% certainty is required, the final decision about the link can be deferred to a human expert.

3. **Data fusion**. After schema mapping and identity resolution the content is harmonized in terms of properties and concept URIs used. However, there is still the problem of dealing with multiple instances of data: the aggregated dataset may contain multiple instances for shared resources such as persons, addresses, and places. The problem is that the property values of these instance may be different: one dataset may have more properties in use than another, or more detailed values, and the values may even be conflicting. For example, the population data

of cities in different language versions of DBpedia differ and one has to decide what value to use.

The final phase in content aggregation is *data fusion* addressing this problem. The goal of the data fusion phase is to create one canonical record for multiple copies of data. The challenge is how to select properties and values. Several strategies are possible here, usually based on metadata considerations, and depending on the case. For example, some data sources may be considered more trustworthy than others, latest updated value may be preferred (e.g., for city population), most frequent or average value is selected, and so on.

Data fusion as well as data maintenance in general is based on *provenance (meta)data* about the metadata. Provenance data tells the origin, history, usage, and other similar aspects of objects, and is widely used in memory organizations for, e.g., documenting the context in which an object was used or created, the conservation processes of an objects, and its exhibition history. Storing provenance metadata about metadata is essential in portals, for example, to identifiy what is the source of the metadata and when was it created.

Provenance metadata of linked data triples is usually represented by associating an additional URI to each triple $< S, P, O >$ making them quadruples called *quads* $< S, P, O, G >$, where G is a URI resource. Provenance data can then be represented in a straightforward manner by triples or quads with G as the subject. Triples with the same G can be considered as RDF graphs of their own and are therefore often called *named graphs*. Named graphs make linked data modular.

7.4   QUALITY OF LINKED DATA

A challenge of increasing importance in Linked Data is data quality. LD is produced by different parties based on heterogeneous data sources. Much of the content, such as Wikipedia, is created by individual volunteers with only modest quality control. The metadata is in many cases produced automatically, often from unstructured data and without guarantees for high quality output. The LD services on the Web are often based on demonstrational software packages and open source projects.

Data quality can be defined generally as fitness for use. Therefore, quality of data not only depends on the data itself but also its user and situation in which the data is employed; the quality of a dataset can be high for one purpose but low for another. At the same time, the notion of quality is multi-faceted involving dimensions such as accuracy and timeliness. For assessing data quality, one has to decide on *quality indicators* used for identifying quality issues along the facets (e.g., how long ago the data was updated), and *assessment metrics* for measuring them. Below, facets of quality are discussed concerning the original primary data, metadata describing it, and the publishing services.

7.4.1   DATA QUALITY OF PRIMARY SOURCES

Some primary data quality issues are listed in Table 7.3. Regarding accuracy, there have been concerns about various Web 2.0 data sources produced by non-experts. Timeliness problems occur due to changes in the underlying real world or our knowledge about it. For example, from 1930 until 2006

Pluto was considered a planet in our solar system but not anymore, population in cities varies in time, presidents of countries change after elections, a painting turns out to be a fake, and so on. Many datasets are incomplete, such as historical gazetteers of places or persons, and often include irrelevant data that is not usable in a particular application.

Table 7.3: Quality issues of LD on a primary data level	
Accuracy	Is the data accurate and correct?
Timeliness	Does the dataset contain outdated information?
Completeness	Does the dataset cover the area of knowledge needed for the application?
Boundedness	Does the data include extra irrelevant data that pollutes the actual data in focus?

7.4.2 METADATA QUALITY

On a metadata level, additional quality issues arise as listed in Table 7.4.

Table 7.4: Quality issues of LD on a metadata level	
Accuracy	How accurately can the metadata schema and annotations describe the content?
Consistency	Does the data contradict itself?
Timeliness	Metadata change in time when data changes.
Provenance and attribution	Is it possible to attribute a piece of data to the authority responsible for it in order to evaluate the trustworthiness of the content? Is it possible to find out when the data was produced?
Licensing	Is the license for using data clear? Is the information available for both humans and machines?
Modeling granularity	Does the metadata model capture enough information to be useful?
Connectedness	Do combined datasets join at the right points?
Homogeneity and isomorphism	Are the combined datasets modeled in a compatible way?

There are validators for RDF(S) data for finding out syntactic errors in RDF and inspecting the underlying graph structure[8]. Validation can be done also on a higher level. For example, if an RDFS schema requires that the range of a predicate is an instance of a class, and a literal value is encountered, an error can be reported. Inconsistencies for OWL ontologies can be detected by OWL reasoners, but debugging out what is actually wrong in an ontology can be challenging.

Assuming a closed world, constraints in a metadata schema can be used for checking semantic validity of a given dataset, e.g., that the value of an element is indeed an instance of a class in a given range. The open world assumption often used on the Semantic Web is more challenging in this respect. For example, missing information is not considered an error but as a goal for the machine to reason.

[8]Cf. e.g., http://www.w3.org/RDF/Validator/

7.4.3 QUALITY OF LINKED DATA SERVICES

Finally, new quality issues arise on a service level, as listed in Table 7.5.

Table 7.5: Quality issues on an LD service level

Efficiency	How much memory and time is needed for processing queries of different kind? For example, it is easy to make SPARQL queries that may return as an answer very large result sets, e.g., the whole database. The same problem occurs with autocompletion as the end-user types in a keyword letter by letter.
Scalabity to multiple users	How scalable is the service in terms of simultaneous users? Does the service scale up dynamically by using additional servers in situations of heavy usage?
Sustainability	Is there a credible organizational process for maintaining the data, metadata, and service?

7.5 BIBLIOGRAPHICAL AND HISTORICAL NOTES

Ontology development 101[9] is a gentle tutorial of ontology modeling including learning materials using the Protégé editor. Ontology modeling and construction using RDF(S) and OWL is discussed in more depth in [2] and [139]. DOLCE methodology for organizing ontologies is discussed in [44]. Ontology learning, also called *ontology extraction*, ontology generation, or ontology acquisition, is a subarea of information extraction. Information extraction is also needed for automatic annotation when transforming legacy literal data into URI references and RDF structures. A recent survey of research and development in this area is presented in [157], and several collections of papers covering the area have been edited [23, 24]. An overview and comparison of approaches to transforming relational databases into RDF is presented in [128].

Transformation of various thesauri into SKOS ontologies without human intervention is discussed in [150]. The thesaurus-to-ontology transformation model presented in this chapter was presented in [68].

Ontology alignment (matching) is discussed, e.g., in [37]. This area has an active research community organizing a workshop series and ontology alignment competitions[10].

Several frameworks for building applications on top of linked data have been developed, such as the Linked Data Integration Framework (LDIF) [131], Information Workbench [49], and SAHA-HAKO [89]. Google Refine[11] is a tool for cleaning up messy data, transforming it to different formats, and for data linking. It is extended by RDF Refine[12], a tool for reconciling, interlinking, and exporting RDF data, and in Linked Media Framework[13], another framework for publishing linked data, building search applications, and using SKOS for information extraction.

---

[9]http://protege.stanford.edu/publications/ontology_development/ontology101.html
[10]http://oaei.ontologymatching.org/
[11]http://code.google.com/p/google-refine/
[12]http://refine.deri.ie/
[13]http://code.google.com/p/lmf/

Quality of Linked Data is an important but fairly new research topic [18, 106, 123, 141]. The criteria presented in the chapter were inspired by criteria listings available on the Web[14]. Links to research and works related to the topic are listed in the Quality Web Data site[15].

---

[14]See   http://sourceforge.net/apps/mediawiki/trdf/index.php?title=Quality_Criteria_for_Linked_
Data_sources, http://answers.semanticweb.com/questions/1072/quality-indicators-for-linked-data-
datasets
[15]http://qualitywebdata.org/

CHAPTER 8

Semantic Services for Human and Machine Users

The goal of semantic information portals for cultural heritage is to provide the end-user with intelligent services for finding the right information and learning based on her own preferences and the context of using the system. In this chapter, some possibilities of providing the end-users with intelligent services using semantically annotated metadata are explored. First, some key notions of classical Information Retrieval (IR) are presented as a baseline. Then ideas of semantic search for enhancing IR using ontologies and linked data are discussed. With semantic models of the end-user and search context, searching can be customized according to the user and the dynamic use environment. After this, three semantic search paradigms are introduced: semantic autocompletion, faceted search, and relational search. The closely related area of semantic recommender systems is also covered. In conclusion, ways of visualizing semantic structures are presented, and cross-portal re-use of semantic services is discussed.

8.1 CLASSICAL INFORMATION RETRIEVAL

Traditional *Information Retrieval* (IR) focuses on text search, based on finding occurrences of words in documents. The end-user formulates his information need as a query, e.g., as a set of words or a Boolean query expression, that is then input into the search system containing a set of documents. Documents that are considered relevant for the query are returned as hits in the *result set*, and irrelevant documents should not be part of the result.

An IR system is typically evaluated in terms of two key parameters: precision and recall. Let *Rel* be the set of relevant documents and *Ret* the set of retrieved documents for a query. Then *Precision* is defined as the fraction of correctly retrieved relevant documents ($Rel \cap Ret$) in the result set:

$$Precision = |(Rel \cap Ret)|/|Ret| \qquad (8.1)$$

Recall is the proportion of the relevant documents retrieved:

$$Recall = |(Rel \cap Ret)|/|Rel| \qquad (8.2)$$

It is easy to make search precise at the cost of recall and vice versa. For example, retrieving simply all documents for all queries would give us 100% recall, but the precision would then of course be very low. A combination of precision and recall values is therefore often used as a measure for IR quality, such as their harmonic mean, i.e., the *F-measure F*:

$$F = 2 * Precision * Recall / (Precision + Recall) \qquad (8.3)$$

The result set is often very large. Then the results need to be ordered according to some measure of relevance. Many search paradigms and algorithms provide means for not only finding relevant documents for a query but also determining their relative measure of relevance, and order the results accordingly. A widely used method is the *vector space model* [129] where the terms occurring in the document set are considered as dimensions. If the document set contains n terms, then each document d_i can be represented as a vector

$$d_i = < w_1, w_2, \ldots, w_n > \qquad (8.4)$$

in the n-dimensional term space where weight w_j, $j = 1, \ldots, n$, tells the number of occurrences of the j:th term in the document. A query in this model is a set of query terms and can also be represented as a similar vector. The relevance of a document w.r.t. a query can then be determined using a similarity measure between the query and the documents. A widely used way to measure similarity is to compare the angles of the vectors by computing the cosine of their deviation (that can be calculated effectively).

Another widely used technique is the *tf/idf method* where documents with more occurrences of query terms (term frequency *tf*) are considered more relevant than others, but the effect is compensated for query terms that occur frequently over the document set (inverse document frequency *idf*). On the Web, the linking structure of documents provides additional information for determining document relevance. A famous example of utilizing this is Google's *PageRank algorithm* that considers pages with more incoming links from respected pages more relevant than others.

Traditional IR deals with text documents. With non-textual cultural documents, such as paintings, photographs, and videos, search can be based on the data itself leading to *content-based information retrieval* methods (CBIR) [122] and *multimedia information retrieval* (MIR) [90] as complementary techniques. Here the idea is to utilize in IR actual document features at a data level, such as color, texture, and shape in images. For example, an image of Abraham Lincoln could be used as a query for finding other pictures of him, or a piece of music could be searched for by humming it. Bridging the *"semantic gap"* between low level image and multimedia features and semantic annotations is an important but challenging research theme [61]. Usually, however, searching non-textual documents is based on searching metadata about the documents. Here Semantic Web technologies focusing on metadata and ontologies are playing an important role.

A fundamental problem in classical IR methods, such as the vector space model and tf/idf, is that they are based on ambiguous literal words, not their underlying meanings. For example, synonyms and homonyms occur in the documents and the query, which makes the search ambiguous and lowers both the precision and recall. Semantic search addresses these problems by founding search on concepts instead of their ambiguous literal expressions.

It should be noted, however, that traditional IR techniques can be adapted and applied to semantic search, too, providing it with a scalable technological basis. For example, the vector space model can be used more semantically if documents and queries are represented as concept (URI) vectors. If ambiguous terms in documents are disambiguated into their correct meanings during the content annotation and indexing phase, and query terms before querying, then higher precision and recall values are possible than without disambiguation.

Below, approaches to semantic search are briefly introduced.

8.2   SEMANTIC CONCEPT-BASED SEARCH

Semantic search refers to concept-based search techniques where additional semantic information, possibly external to the dataset at hand, is utilized in order to make search more "intelligent" than with classical IR techniques. Such additional external information may concern the ontological properties of the data to be searched, the end-user's preferences, profile, and/or actions, or the spatio-temporal or social context where searching is performed. The goal of utilizing additional semantic information is to understand the user's information need (query) more deeply in a context, and in this way to determine the relevance of search objects more accurately.

8.2.1   HANDLING SYNONYMS

Synonyms lower the recall of IR. For example, a document about "students" may not be found using the query term "pupils." Document annotations on the Semantic Web are based on concept URIs, not on their literal labels for human consumption. Concepts typically have alternative synonymous names in different languages as properties. This makes it easy to find documents where any of the synonyms referring to the same concept occurs, including expressions in different languages when multilingual ontologies are available.

The problem of synonyms is particularly difficult when dealing with names. Firstly, names can be written using different syntactic conventions, e.g., "Pyotr Ilyich Tchaikovsky" vs. "Tchaikovsky, P.." Second, the names may be written using different transliteration systems, e.g., "P. Chaykovsky" vs. "P. Tchaikovsky." Third, a person may have different names during his/her life (due to marriage, for example), use artistic pseudo-names or pen names by themselves (e.g., Samuel Clemens published books with the name Mark Twain), and are given nicknames by others (e.g., Bruce Springsteen, the musician, is known as the Boss).

In semantic ontologies it is also possible to associate concepts and words with their *antonyms*, e.g., "death" to "birth," as in WordNet. In some cases documents about antonyms concepts are related and could be returned as search hits.

8.2.2   HOMONYMS AND SEMANTIC DISAMBIGUATION

Homonyms, on the other hand, result in lower precision. For example, a document about "pupils" may be about students or about the organ eye, or about both. When searching for documents about the person "Paris Hilton," documents concerning the capital of France and hotels may be falsely included in the results. A challenging problem in search is how to deal with words having multiple meanings.

In order to retrieve the correct documents, semantic disambiguation of literal word meanings is needed 1) when indexing data and 2) also when interpreting the query. When indexing data, disambiguation can be done manually by using, e.g., keywords taken from an ontology, or automatically based on the context of the word. For example, the vicinity of the word "eye" suggests that "pupil" may refer to the organ rather than to student, but not necessarily. Semantic disambiguation is a widely studied area of research in Information Extraction and Natural Language Processing [3, 102].

Queries are typically short and little textual context is available for disambiguation. However, in any case, ontologies make it possible to identify multiple meanings and help the end-user in making the choice. For example, semantic autocompletion can be used to give the end-user the choices to select from, for example, "pupil (student)" or "pupil (part of eye)."

8.2.3 QUERY AND DOCUMENT EXPANSION

Further enhancements of search are possible by using ontology-based *query expansion* or *document expansion*. In ontology-based query (or document) expansion, semantic relations in ontologies are used to extend the query vector (or document vectors) with additional concepts in order to describe them more accurately. For example, when searching for "birds" the query can be expanded along a biological ontology into a disjunctive query about "swans," "eagles," etc. In the same vein, if the concept "Nordic countries" occurs in a document, then the concepts "Denmark," "Finland," "Island," "Sweden," and "Norway,"[1] could be added in the document annotation based on the part-of relations in a place ontology. As a result, the document would be found when querying with individual country names.

Query and document expansion are used to raise the recall in IR, and are often quite successful especially when using semantically annotated contents and disambiguated queries. However, query and document expansion are not a panacea for IR. Too much expansion easily results in lower precision, and it is not always easy to decide how and how much to expand. For example, when querying with "Nordic countries" the user's information need may actually be related to Nordic countries only as a whole, e.g., to common agreements or organizations between the countries in the Nordic community, not to individual countries. It is also likely that expansion along the part-of relation should not go down to cities and villages in the countries. Such problems of query/document expansion are obviously harder when expanding documents and queries in a literal space where ambiguities due to synonymy and homonymy introduce additional difficulties.

---

[1]Possibly with their associated territories, such as the Faroe Islands and Greenland.

8.3   SEMANTIC AUTOCOMPLETION

A problem related to search is how to select terms in a query. Only terms that are relevant to the domain and content available are meaningful; other queries result in frustrating "no hits" answers. *Autocompletion* has become a popular way to find meaningful search terms after Google Suggest[2] was released. Before that, autocompletion was already widely used, e.g., in programming tools.

The idea of autocompletion is simple. Since the computer knows the meaningful term space for formulating queries, it can suggest to the user after each input character the possible terms matching the input string thus far. Suggestions can be ordered, e.g., alphabetically or based on their popularity. This speeds up typing a lot. Autocompletion is especially useful in mobile devices where inputting characters is usually tedious.

The general idea of *semantic autocompletion* [62] is to try to autocomplete the input string not only on syntactic grounds but based on semantic ontologies. This idea can be applied, e.g., to disambiguating query terms into different options from which the end-user can select the intended meaning. For example, the ambiguous search term "Nokia" matches in the semantic portal [63] with 1) the mobile phone company, 2) with the city of Nokia, 3) and with some other organizations whose names contain the string "Nokia." Furthermore, these resources can be used in different roles regarding the underlying museum collection data, such as in the "manufacturer" role (organization resources) and in the roles "place of manufacturing" and "place of usage" (the city resource). Semantic autocompletion can be used to expose these interpretations of "Nokia" to the end-user to select from, and after this a correct search with the intended meaning can be accurately performed. There are also other possibilities for doing autocompletion semantically using ontologies, such as autocompletion across different languages. A form of semantic autocompletion is to expand the input string not only to search terms but all the way down to the actual search results.

8.4   FACETED SEMANTIC SEARCH AND BROWSING

Autocompletion is a way to provide the end-user with a querying vocabulary. The traditional way to providing the vocabulary is to use an index as in printed books. When using ontologies the index can be provided not only as an alphabetical list but as a semantic structure, such as a classification or a subject heading category tree classifying the search objects along a dimension called *facet*. By selecting a category in the facet, additional choices can be shown or related documents can be retrieved. This approach is used, e.g., in the Yahoo! directory and in the dmoz Open Directory Service[3].

This idea of using facets has been generalized into a search paradigm called *faceted search*, also known as *view-based search*, or *dynamic hierarchies*. In faceted search a number of facets is exposed to the user classifying the content along semantically orthogonal dimensions. For example, an artifact collection could have facets such as object type, creator, place of creation, time of usage, etc. The facets are exposed to the end-user in order to 1) provide her with the right query vocabulary in

---

[2]http://www.google.com/webhp?complete=1&hl=en
[3]http://dmoz.org

terms of intuitive hierarchies, and 2) for presenting the repository contents and search results along facet categories. A query is formulated by selecting categories in the facets in an arbitrary order. After each selection documents matching all selections at the same time are returned and a hit count number is shown for each category indicating the number of hits if the category is selected next. This information efficiently guides the user in making selections and eliminates completely queries leading to "no hits" dead-ends.

Faceted search is useful especially in situations, where the user cannot formulate his information needs in accurate terms, but rather wants to have a look and learn about an area of interest. According to user tests, Google-like keyword search interface is usually preferred if the user knows exactly what she is looking for (e.g., a particular object in a collection) and is capable of expressing her information need accurately [36].

Faceted search fits well and has been integrated with the idea of hierarchical ontologies and the Semantic Web [65]. The facets can often be constructed algorithmically from a set of underlying ontologies that are used as a basis for annotating search items. Furthermore, the mapping of search items onto search facets can be defined using, e.g., SPARQL queries or logic rules, which facilitates more intelligent semantic search of indirectly related items.

Faceted search can be extended in different ways. For example: If the number of hits is large they can be ordered according to a relevance measure. Facets to be exposed to the end-user may be determined dynamically based on earlier selections and/or the current result set, which may be helpful in focusing next search selections effectively on a facet level. It is possible to keep all paths of search selections open and visualize them to the end-user like in breadth-first search. Facets can be visualized in different ways using a map, a timeline, or other graphical representations, such as an image of a complex construction or process description. A text search box can be implemented and visualized as a filtering facet, too.

Ontology hierarchies provide a natural starting point for designing facets. However, in many cases, ontologies are not intuitive enough as facets to the end-user, or do not provide a perspective needed in search. Then separate facets can be constructed by hand and aligned with the ontologies used in annotations.

8.5   SEMANTIC BROWSING AND RECOMMENDING

The Web is used in two principal ways: by searching and by browsing. Searching finds links to directly relevant documents based on a query. Browsing means information finding by following associative links on the Web pages found.

Linked data facilitates *semantic browsing* in addition to semantic search. Faceted search is actually already a kind of combination of searching and browsing because search is based on selecting links on the facets. However, in semantic browsing the general idea is not to constrain the result set but rather to *expand* it by trying to find objects of potential interest outside of the hit list. The idea is to support browsing documents through associative recommendation links that are created based on the underlying linked metadata and ontologies, not on hardwired anchor links encoded by humans in the HTML pages. Semantic recommendations are usually based on semantic criteria that are different from and complementary to those used in search. For example, statistical information about the portal use of other similar users provides one possible basis for this.

A simple form of semantic browsers are RDF browsers and tabulators for exploring the Web of Data through the arcs in the RDF graph, as discussed in Section 3.2. A more developed related idea is *recommender systems* [25, 74], also known as *recommendation systems*. A recommender system provides the end-user with recommendations of items based on her personal interests and the content at hand. For example, when looking for a book on a portal, links to other similar books of potential interest can be recommended to the end-user, or links to biographies of the author.

Recommender systems fall into three major categories.

1. **Collaborative filtering** systems [56] are based on creating and utilizing user profiles of end-users. Such descriptions can be used for identifying user groups of similar interests. When a member in a group makes a selection, e.g., takes a look at a book or buys it, then the same book can be recommended to the other members of the group. Collaborative filtering suffers from the *cold start problem*: it is not possible to make recommendations before the user(s) have already made several choices for the statistics.

2. **Content-based recommending** systems make use of the characteristics of the objects recommended, such as the genre, topic, and author of books. Recommendations can then be made without any statistics. For example, other books with similar topics can be recommended.

3. **Knowledge-based recommending** is often needed when creating recommendations in unique decision making situations where user statistics or structured data about the objects alone are not enough for determining a recommendation. Knowledge-based recommending can help, for example, in selecting a digital camera to buy that matches the user's personal preferences and goals. The recommending process may be interactive.

All these forms of recommending are worth considering in semantic CH portals, including hybrid recommender systems where different approaches are combined in one way or another. In particular, linked data structures can be used as a semantically rich basis for content-based and knowledge-based recommendations. For example, logic rules on top of an RDF knowledge base can be used for creating semantic recommendation links and, at the same time, *explanations* telling the end-user why the recommendation link was selected in this context [153]. In [5, 7] explanations for recommended artworks are obtained based on a user profile of interest, features of the artworks, and previous selections of artworks for viewing. In [77, 85] ontological models of narrative stories and processes in the society, such as fishing or slash-and-burn farming, were used as a basis for creating recommendation links between cultural resources. Still another approach to create recommendation links with explanations is to use similarity measures of event-based annotations [125].

Ontologies can also be used for representing user preferences and profiles. Same concepts as in content annotations can be used for interoperability.

8.6   RELATIONAL SEARCH

The goal of *relational search*, also known as *association discovery*, is to try to discover serendipitous semantic associations between two (or more) given content items [64, 132, 134]. This is a different form of semantic recommending where items of interest from one given item to any other item are searched for.

Relational search makes it possible for the end-user to formulate queries such as "How is *X* related to *Y*" by selecting the end-point resources. The search result is a set of semantic connection paths between *X* and *Y*, typically with some kind of explanation about the relationship. For example, in Figure 8.1 the user has specified two historical persons, the French emperor Napoleon I (1769–1821) and the Finnish artist Akseli Gallen-Kallela (1865–1931), and the system discovered an association chain between the persons based on "patronOf," "teacherOf," "knows," and "studentOf" properties in the RDF graph. The underlying knowledge base is based on the ULAN actor gazetteer in RDF form that is also available in an interactive graph form for browsing. Autocompletion is used for finding the right query resources. RelFinder[4] is a general tool for extracting and visualizing relationships between resources in an RDF graph, and exploring them interactively.

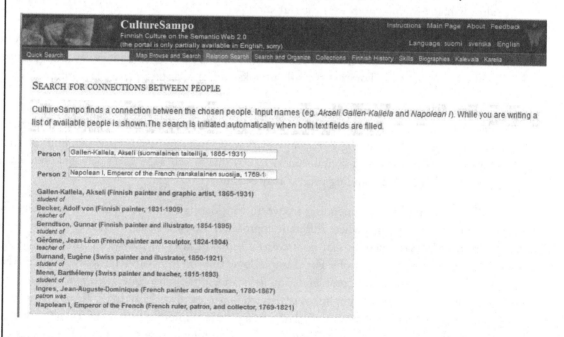

Figure 8.1: An example of relational search in [64] using the ULAN vocabulary and database.

---

[4]http://www.visualdataweb.org/relfinder.php

8.7   VISUALIZATION AND MASH-UPS

Visualization is an important aspect of the Semantic Web, dealing with semantically complicated and interlinked contents. Visualizations provide tools for humans to understand better the underlying data, aggregate and filter out features and trends in it, analyze data, and in this way solve problems in Digital Humanities research. Easy to grasp graphical illustrations of CH are particularly essential for layman users.

Visualizing linked data structures may concern relationships between datasets, ontologies, metadata in a dataset, and search results. In this section some possibilities for visualizing linked data are explored.

8.7.1   VISUALIZING DATASET CLOUDS

Thousands of datasets have been linked with each other through ontology and data mappings. Linked datasets and their mappings form their own structures that are in many cases illustrative to inspect through graphical representations.

The most famous visualization in this category is the Linked Open Data cloud depicted in Figure 3.1. Here datasets in different domains are depicted as colored bubbles and mappings between them as connecting arcs. Another visualization[5] of the LOD cloud has been created using the Protovis tool[6]. Colors are in this case used to indicate different ratings of the datasets in the CKAN registry.

8.7.2   VISUALIZING ONTOLOGIES

Linked data browsers typically provide links to directly related resources based on RDF properties. When a larger view of the underlying network is needed, the whole graph or part of it can be visualized for the user. This is often useful when inspecting hierarchical ontologies and other knowledge structures [79].

Visualizations can be useful for illustrating to the user how the concepts are related with each other. The baseline is to provide hierarchical views to ontologies like folders and files in a file system, but there are other ways, too, such as 2D and 3D graphs, zoomable visualizations, and space filling techniques where the screen space is divided into subareas corresponding to subconcepts of a resource. Such visualizations are typically used in ontology editors.

Network-based representations can be difficult to understand and use when the number of nodes and relations between them is large. To alleviate the problem, one may visualize only the vicinity of a resource in focus or emphasize it graphically. Some representations of knowledge, such as arithmetical formulas and embedded quantifications in logic may be more difficult to grasp as graphs than in conventional linear form.

---

[5]http://inkdroid.org/lod-graph/
[6]http://mbostock.github.com/protovis/

8.7.3   VISUALIZING METADATA

All Linked Data browsers have a way to present the properties and values of a resource, such as a collection item in a portal. Depending on the application case some properties can be filtered out and the remaining data be presented and formatted in a desired form. Fresnel[7] is an example of a vocabulary by which such visualizations can be specified.

In many cases there is the possibility of projecting and visualizing resources of interest on graphical illustrations of the underlying real world. In the CH domain, especially maps and timelines are often used for this purpose. However, photographs, blueprints, and other images can also be used for this.

Maps are useful in both searching content and in visualizing the results. A widely used way of utilizing maps in portals is to use Google or Yahoo! maps or similar services for mash-ups. Such map APIs are easy to use for showing data resources on maps based on either their direct coordinate data or references to a place ontology. By clicking on an item on the map, more information about it can be shown to the end-user with further links. Often items are related to places in several ways. For example, a painting about Paris could have been painted in London and is situated in Madrid. Different relations can be visualized on the maps by, e.g., different symbols or colors.

In the CH domain, *historical maps* are of interest of their own. They may depict old place names, borders, and geographical features not available anymore in contemporary maps. An approach to visualize historical changes is presented in [80, 81]. Here old maps are laid semi-transparently on top of the contemporary maps and satellite images of Google Maps. This makes it possible to view old and new place names on maps at the same time, and visualize changes of geographical features (roads, villages, etc.) in history. Changes can be quite dramatic when a region is annexed to another country due to war, for example.

Another important dimension for visualizing cultural content is time. A standard approach for temporal visualization is to project search objects on a timeline to visualize trends or developments in time. Ready to use software packages to do this are available, such as the Simile Timeline[8]. For example, in [72] pictorial collection items (paintings, photographs, etc.) related to the concept "beard" were projected on a timeline to inspect the changes of beard fashion in the 19th century. A timeline can be used in the same vein as a map. By clicking on an item projected on a timeline, more information about the item can be shown to the user. A timeline can also be used as a facet for querying.

Above, maps and timelines were used for presenting resources as separate points. It is also possible to visualize relations. For example, in the Spatial History Project[9] and Mapping the Republic of Letters Project[10] Voltaire's (and other authors') correspondences have been visualized on maps as lines between the place where a letter was written and the place where it was sent. The underlying research question here is to determine the national and global extent of Voltaire's correspondence

---

[7]http://www.w3.org/2005/04/fresnel-info/manual/
[8]http://www.simile-widgets.org/timeline/
[9]http://www.stanford.edu/group/spatialhistory/cgi-bin/site/index.php
[10]http://republicofletters.stanford.edu

networks. Similar kinds of visualizations were presented in [83], where arrows on the map show connections between the places where artifacts of collections were manufactured and actually used. The thickness of the arrows indicated the number of artifacts sharing history in this way. This kind of visualization shows the local and global flow of artifacts (import and export) in an intuitive way. A similar kind of visualization was also used to visualize where people in the ULAN registry were born and died, showing local and global flows of CH celebrities to major cultural cities, such as New York.

Connections between two places can be chained further into routes on maps, such as guided tours for contemporary cultural tourists or visualizations of famous explorations and other historical trips. Portal contents can be projected along such paths.

8.7.4 VISUALIZING SEARCH RESULTS

In addition to visualizing the contents of an RDF repository, visualizations are useful for inspecting search results from different perspectives. For example, the result set can be presented not only as a list of hits in a relevance order, but also clustered along ontological distinctions, e.g., presented by content type, according to related places, or along a timeline. In faceted search, for example, the facets provide natural dimensions and categorizations for such purposes. It is possible to classify materials at the same time along two or even more dimensions. For example, in [100] artifacts manufactured in Japan and used in another country where projected on a timeline showing a change from low-tech to high-tech products in imports from Japan in the 20th century.

Another useful approach is to use business graphics such as histograms, pies, and diagrams for inspecting and analyzing the results [104]. There are packages available for this, such as sgvizler[11] for creating Google Chart graphics on top of SPARQL endpoints.

Also augmented reality visualizations for CH collections can be used. For example, using systems such as Layar[12] and a mobile phone with a camera and a compass, CH contents in a particular direction seen from the current position can be visualized on the camera image.

8.8 PERSONALIZATION AND CONTEXT AWARENESS

On many occasions the functioning of a semantic portal should not be static but adapt dynamically according to the personal interests and abilities of the end-user and the context of usage, such as time and location [135] or social context. Visitors in semantic cultural portals, like in physical museums, are usually not interested in everything found in the underlying collections, and would like to get selected information at different levels of detail. An important aspect of a semantic cultural portal is then adaptation of the portal to different personal information needs and interests.

Personalization needs can be related in principle to any metadata type. The user may be interested, for example, in particular object types (say artworks), subject matters (say nature and animals), or creators (say Impressionists).

[11]http://code.google.com/p/sgvizler/
[12]http://www.layar.com/

Personalization is based on a user profile or model. Such a model can be created either *explicitly* by asking categories of interest from the user by a questionnaire. However, eliciting such descriptions can be difficult. It is also possible to try to create profiles in an *implicit* manner by monitoring end-user's behavior, choices, or using like/dislike voting. For example, in [5, 7] personalization is based on metadata obtained by asking the user to rate pieces of artworks she has viewed.

Location-based adaptability is widely used in mobile phone applications. Based on GPS or other locating services, objects or services in the vicinity of the user can be retrieved. In rich linked data, there can be different kinds of relations between a CH object and a place that can be explained to the end-user. For example, collection artifacts may have been manufactured or used at a place, a photograph or a painting may depict the place, a person may have been born or died at a place, a poem has been collected there, and so on.

Also, time is an important parameter for contextualizing portal services. For example, it may not be wise to recommend the end-user to visit a ruin during winter, if it is covered with snow, or direct her to a museum on a Monday when it happens to be closed.

The user's inherent properties, such as age (child, grown-up, elderly), sex, education, linguistic skills, and disabilities regarding seeing, hearing, and moving can be important parameters in personalization. Also, available means for transportation may be of importance when looking for points of interest.

There are ontologies, such as GUMO [54], designed for representing aspects of personalization and context in a machine interpretable manner.

8.9   CROSS-PORTAL RE-USE OF CONTENT

Portal contents can be re-used in other Web applications and portals based on Semantic Web standards. Re-using semantic content in this way is a kind of generalization of the idea of "multi-channel publication" of XML, where a single syntactic structure can be rendered in different ways. In a similar vein, semantic metadata can be re-used without modifying it through *multi-application publication*. The standard way to facilitate this is to use linked data APIs discussed in Section 3.2.

One possibility to facilitate cross-portal re-use is to simply download RDF graphs, merge them, and provide services to end-users based on the extended knowledge base. For example, the learning object video portal presented in [78] is able to provide recommendation links to a cultural museum collection portal in this way.

Another way of re-using content is to keep the portals separate and use Web services, such as SPARQL endpoints or recommendation link servers [154]. In addition to SPARQL querying, traditional Web Services based on SOAP and WSDL or light-weight REST APIs on HTTP can be used. With a little additional programming work, APIs can be wrapped into AJAX widgets that can be used in other portals' HTML code in a similar vein as Google AdSense[13] is used for publishing advertisements on external Web pages. For example, if an article of a newspaper site concerns the

---

[13]http://www.google.com/adsense/

history of skating, then such a widget can query and show dynamically images and semantic links to skates and related objects in a museum collection portal [95].

8.10   BIBLIOGRAPHICAL AND HISTORICAL NOTES

Lots of textbooks are available concerning IR, such as [11, 101]. An extensive review of content-based multimedia IR research is presented in [90].

The faceted search paradigm is based on *facet analysis* [103], a classification scheme introduced in information sciences by S. R. Ranganathan in the 1930's. The idea of faceted search [52] was invented and developed independently by several research groups, and is also called view-based search [117] and dynamic taxonomies [127]. The idea was integrated with Semantic Web ontologies in [65] and with fuzzy logic in [60]. In [142] a card sorting approach is presented for specifying and using intuitive end-user facets independently from indexing ontologies.

Since Google Suggest, autocompletion has become a standard technique on the Web for helping the end-user in indexing contents and in formulating meaningful queries. Autocompletion functionality can be provided easily for external use as Web services.

Semantic recommending [35, 107], relational search, and knowledge discovery [134] are interesting topics of research. A key question there is how to characterize and specify the notion of serendipity in such as way that useful nontrivial semantic relations can be found and explained to end-users.

The Conference on User Modeling, Adaptation, and Personalization (UMAP)[14] is a major international conference series for researchers and practitioners working on systems that adapt to their users and represent information about them for this purpose. The annual International Workshop on Semantic Media Adaptation and Personalization (SMAP) has been organized and proceedings published since 2006 [156]. Personalization from a museum perspective is discussed, e.g., in [7, 20].

A collection of articles related to Semantic Web visualization is contained in [29, 45] and a survey of approaches is presented in [79].

Re-use of semantically interoperable contents is one of the main goals of the Linked (Open) Data movement discussed in this book.

[14]http://www.um.org/

CHAPTER 9

Conclusions

Cultural heritage provides a semantically rich application domain in which useful vocabularies and collection contents are available, and where the organizations are eager to make their content publicly accessible. A major application type in the area has been semantic portals, often aggregating heterogeneous contents of different organizations from distributed collections. This publishing scheme provides cultural organizations with a shared, cost-effective publication channel and the possibility of enriching collaboratively their contents with each other's linked data. Semantic Web and Linked Data technologies provide a promising basis for creating and maintaining such systems.

This book started by formulating and motivating a sustainable "business model" for semantic CH portals, as well as a general architecture and framework for them (Chapters 1 and 2). The model is based on requirements for publishing linked data (Chapter 3).

Next, technological foundations for semantic portals were explained based on the layer-cake model of the Semantic Web: metadata models for Web publication, cataloging, content harmonization, and data harvesting (Chapter 4), domain ontologies for filling in values in metadata templates (Chapter 5), and logic rules for enriching contents automatically by reasoning (Chapter 6).

After this, content creation processes for vocabularies, ontologies, and linked data were in focus including data quality issues (Chapter 7). For the end-user, new kinds of intelligent semantic search and recommending services as well as ways of analyzing and visualizing contents can be provided (Chapter 8).

It is easy to envision that the development is leading toward larger semantic CH portals, since larger and larger linked datasets with better and better quality are being published. Such datasets are crossing geographical, cultural, and linguistic barriers of content providers in different countries. Another direction of progress is enriching the collections by end-user created content and tagging. In the end, cultural heritage, as a social construct, is created by us all. At the same time, collection data is being provided more and more often in personalized ways and in a spatio-temporal context thorough mobile devices.

The big challenge in creating CH portals is how to make heterogeneous contents semantically interoperable and mutually interlinked. It is fairly easy to make applications if their basis, the data, is good and well-linked. A major practical barrier for this is that current legacy cataloging systems in memory organizations do not support creation of ontology-based annotations. If semantic annotations cannot be created in memory organizations already when cataloging content, then costly manual work is needed later on when transforming and disambiguating literal metadata for shared portal use. A way to support ontology-based cataloging is to create a public domain specific content infrastructure of ontologies, based on domain independent Semantic Web standards, and to make it operational by ontology and annotation services than can be connected easily with cataloging systems [69]. Albert Einstein's aphorism applies well to solving interoperability problems of cultural heritage contents:

Intellectuals solve problems, geniuses prevent them.

Bibliography

[1] J. Aitchison, A. Gilchrist, and D. Bawden. *Thesaurus construction and use: a practical manual*. Europa Publications, London, 2000. Cited on page(s) 60, 88

[2] David Allemang and Jim Hendler. *Semantic Web for the working Ontologiest*. Morgan Kaufmann, San Francisco, 2008. Cited on page(s) 77, 104

[3] David Allemang and Jim Hendler. *Speech and Language Processing, 2nd edition*. Prentice-Hall, Upper Saddle River, New Jersey, 2008. Cited on page(s) 110

[4] Grigoris Antoniu and Frank van Harmelen. *Semantic Web Primer*. The MIT Press, Cambridge, Massachusetts, 2008. Cited on page(s) 77, 83, 86

[5] L. Aroyo, R. Brussee, L. Rutledge, P. Gorgels, N. Stash, and Y. Wang. Personalized museum experience: The Rijksmuseum use case. In J. Trant and D. Bearman, editors, *Museums and the Web 2007: Proceedings*, pages 137–144, 2007. Cited on page(s) 38, 113, 118

[6] L. Aroyo, E. Hyvönen, and J. van Ossenbruggen, editors. *Cultural Heritage on the Semantic Web. Workshop Proceedings. The 6th International Semantic Web Conference and the 2nd Asian Semantic Web Conference, Busan, Korea*. ISWC + ASWC, 2007. http://www.cs.vu.nl/~laroyo/CH-SW.html. Cited on page(s) 20, 132

[7] L. Aroyo, N. Stash, Y. Wang, P. Gorgels, and L. Rutledge. CHIP demonstrator: Semantics-driven recommendations and museum tour generation. In *Proceedings of ISWC 2007 + ASWC 2007, Busan, Korea*, pages 879–886. Springer–Verlag, 2007.
DOI: 10.1007/978-3-540-76298-0_64 Cited on page(s) 20, 38, 113, 118, 119

[8] Murtha Baca, editor. *Introduction to metadata*. Getty Publications, Los Angeles, 2008. Cited on page(s) 35, 55, 126

[9] Murtha Baca and Patricia Harpring, editors. *Categories for description of works of art*. Getty Publications, Los Angeles, 2009. http://www.getty.edu/research/publications/electronic_publications/cdwa/index.html. Cited on page(s) 40

[10] Murtha Baca, Patricia Harpring, Elisa Lanzi, Linda McRae, and Ann Whiteside, editors. *Cataloging Cultural Objects. A guide to describing works and their images*. American Library Association, 2008. Cited on page(s) 40

[11] R. Baeza-Yates and B. Ribeiro-Neto. *Modern Information Retrieval.* Addison-Wesley, 1999. Cited on page(s) 119

[12] Chryssoula Bekiari, Martin Doerr, and Patrick Le Boeuf, editors. *FRBR object-oriented definition and mapping to FRBR(version 1.0).* International Federation of Library Associations and Institutions (IFLA), 2009. http://www.cidoc-crm.org/docs/frbr_oo/frbr_docs/FRBRoo_V1.0_2009_june_.pdf. Cited on page(s) 49

[13] V. R. Benjamins, J. Contreras, M. Blázquez, J.M. Dodero, A. Garcia, E. Navas, F. Hernandez, and C. Wert. Cultural heritage and the semantic web. In *The Semantic Web: Research and Applications*, pages 433–444. Springer–Verlag, 2004. DOI: 10.1007/978-3-540-25956-5_30 Cited on page(s) 20

[14] T. Berners-Lee, M. Fischetti, and M. Dertouzos. *Weaving the Web: The Original Design and Ultimate Destiny of the World Wide Web.* Harper Business, 2000. Cited on page(s) 9

[15] T. Berners-Lee, J. Hendler, and O. Lassila. The semantic web. *Scientific American*, 284(5):34–43, May 2001. DOI: 10.1038/scientificamerican0501-34 Cited on page(s) 10

[16] J. Bhogal, A. Macfarlane, and P. Smith. A review of ontology based query expansion. *Information Processing & Management*, 43(4):866–886, 2007. DOI: 10.1016/j.ipm.2006.09.003 Cited on page(s) 73

[17] C. Bizer, T. Heath, and T. Berners-Lee. Linked data – the story so far. *International Journal on Semantic Web and Information Systems (IJSWIS)*, 2009. DOI: 10.4018/jswis.2009081901 Cited on page(s) 33

[18] Christian Bizer and Richard Cyganiak. Quality-driven information filtering using the WIQA policy framework. *Journal of Semantic Web*, 7(1):1–10, 2009. DOI: 10.1016/j.websem.2008.02.005 Cited on page(s) 105

[19] C. Borgman and S. Siegfriend. Getty's synoname and its cousins: A survey of applications of personal name-matching algorithms. *Journal of the American Society for Information Science and Technology*, 43(7):459–476, 1992. Cited on page(s) 69

[20] J. P. Bowen and S. Filippini-Fantoni. Personalization and the web from a museum perspective. In *Selected Papers from an International Conference Museums and the Web 2004 (MW2004), Arlington, Virginia, USA*. Archieves & Museum Informatics, 2004. http://www.museumsandtheweb.com/mw2004/papers/bowen/bowen.html. Cited on page(s) 119

[21] Ronald Brachman and Hector Levesque. *Knowledge representation and reasoning.* Morgan Kaufmann, San Francisco, 2004. Cited on page(s) 35

[22] Ivan Bratko. *Prolog programming for Artificial Intelligence (4th edition).* Pearson, Harlow, U.K., 2012. Cited on page(s) 86

[23] P. Buitelaar and P. Cimiano, editors. *Ontology Learning from Text: Methods, Evaluation and Applications.* Series information for Frontiers in Artificial Intelligence and Applications. IOS Press, Amsterdam, The Netherlands, 2005. Cited on page(s) 104

[24] P. Buitelaar and P. Cimiano, editors. *Ontology Learning and Population: Bridging the Gap between Text and Knowledge.* Series information for Frontiers in Artificial Intelligence and Applications. IOS Press, Amsterdam, The Netherlands, 2008. Cited on page(s) 104

[25] R. Burke. Knowledge-based recommender systems. In A. Kent, editor, *Encyclopedia of Library and Information Systems*, volume 69. Marcel Dekker, New York, 2000. Cited on page(s) 113

[26] E. Clementine, P. De Felice, and P. van Oosterom. A small set of formal topological relationships suitable for end-user interaction. In A. Abel and B. Chin Ooi, editors, *Advances in databases*, pages 277–295. Springer–Verlag, 1993. Cited on page(s) 73

[27] Erin Coburn, Richard Light, Gordon McKenna, Regine Stein, and Axel Vitzthum, editors. *LIDO – Lightweight Information Describing Objects. Version 1.0.* ICOM-CIDOC Working Group Data Harvesting and Interchange, 2010. Cited on page(s) 49

[28] Nick Crofts, Martin Doerr, Tony Gill, Stephen Stead, and Matthew Stiff (Eds.), editors. *Definition of the CIDOC Conceptual Reference Model, Version 5.0.4.* ICOM/CIDOC Documentation Standards Group (CIDOC CRM Special Interest Group), 2011. http://www.cidoc-crm.org/docs/cidoc_crm_version_5.0.4.pdf. Cited on page(s) 44

[29] Aba-Sah Dadzie and Matthew Rowe. Approaches to visualising linked data: A survey. *Semantic Web – Interoperability, Usability, Applicability*, 1(1–2), 2011. DOI: 10.3233/SW-2011-0037 Cited on page(s) 119

[30] Mathieu d'Aquin and Holger Lewen. Cupboard – a place to expose your ontologies to applications and the community. In *Proceedings of the ESWC 2009*, pages 913–918. Springer–Verlag, 2009. DOI: 10.1007/978-3-642-02121-3_81 Cited on page(s) 77

[31] Mathieu dÁquin and Enrico Motta. Watson, more than a semantic web search engine. *Semantic Web – Interoperability, Usability, Applicability*, 2(1):55–63, 2011. DOI: 10.3233/SW-2011-0031 Cited on page(s) 32

[32] M. Doerr. The CIDOC CRM—an ontological approach to semantic interoperability of metadata. *AI Magazine*, 24(3):75–92, 2003. Cited on page(s) 44

[33] Martin Doerr. Ontologies for cultural heritage. In Staab and Studer [139], pages 463–486. DOI: 10.1007/978-3-540-92673-3_21 Cited on page(s) 44

[34] John Dominque, Dieter Fensel, and James A. Hendler, editors. *Handbook of Semantic Web.* Springer–Verlag, 2011. DOI: 10.1007/978-3-540-92913-0 Cited on page(s) 5, 77

[35] Martin Dzbor, Enrico Motta, and Laurian Gridinoc. Browsing and navigating in semantically rich spaces: Experiences with magpie application. In Staab and Studer [139], pages 687–709. DOI: 10.1007/978-3-540-92673-3 Cited on page(s) 119

[36] J. English, M. Hearst, R. Sinha, K. Swearingen, and K.-P. Lee. Flexible search and navigation using faceted metadata. Technical report, University of Berkeley, School of Information Management and Systems, 2003. Cited on page(s) 112

[37] J. Euzenat and P. Shvaiko. *Ontology Matching*. Springer–Verlag, 2007. Cited on page(s) 93, 104

[38] C. Fellbaum, editor. *WordNet. An electronic lexical database.* The MIT Press, Cambridge, Massachusetts, 2001. Cited on page(s) 59

[39] Tim Finin, Li Ding, Rong Pan, Anupam Joshi, Pranam Kolari, Akshay Java, and Yun Peng. Swoogle: Searching for knowledge on the semantic web. In *In AAAI 05 (intelligent systems demo*, pages 1682–1683. The MIT Press, Cambridge, Massachusetts, 2005. Cited on page(s) 32

[40] D. J. Foskett. Thesaurus. In *Encyclopaedia of Library and Information Science*, volume 30, pages 416–462. Marcel Dekker, New York, 1980. Cited on page(s) 88

[41] J. French, A. Powell, and E. Schulman. Using clustering strategies for creating authority files. *Journal of the American Society for Information Science*, 51(8):774–786, jun 2000. Cited on page(s) 69

[42] Matias Frosterus, Eero Hyvönen, and Mika Wahlroos. Extending ontologies with free keywords in a collaborative annotation environment. In *Proceedings of the ISWC 2011 Workshop Ontologies Come of Age in the Semantic Web (OCAS)*. CEUR Workshop Proceedings, Vol 809, http://ceur-ws.org, 2011. Cited on page(s) 96

[43] C. Galvez and F Moya-Anegon. Approximate personal name-matching through finite-state graphs. *Journal of the American Society for Information Science and Technology*, 58(13):1960–1976, 2007. Cited on page(s) 69

[44] A. Gangemi, N. Guarino, C. Masolo, A. Oltramari, and L. Schneider. Sweetening ontologies with DOLCE. In *Proceedings of the 13th International Conference on Knowledge Engineering and Knowledge Management (EKAW 2002)*. Springer–Verlag, 2002. Cited on page(s) 67, 91, 104

[45] V. Geroimenko and C. Chen, editors. *Visualizing the Semantic Web: XML-based Internet and Information Visualization*. Springer–Verlag, 2002. Cited on page(s) 119

[46] A. J. Gilliland. Setting the stage. In Baca [8], pages 1–19. Cited on page(s) 35

[47] N. Guarino and C. Welty. Evaluating ontological decisions with OntoClean. *Communications of the ACM*, 45(2):61–65, 2002. Cited on page(s) 91

[48] Nicola Guarino, Daniel Oberle, and Steffen Staab. What is an ontology? In Staab and Studer [139], pages 1–17. DOI: 10.1007/978-3-540-92673-3 Cited on page(s) 62

[49] P. Haase, M. Schmidt, and A. Schwarte. The information workbench as a self-service platform for linked data applications. In *Proceedings of the 2nd International Workshop on Consuming Linked Data (COLD 2011)*, 2011. CEUR Workshop Proceedings, Vol 782, http://ceur-ws.org. Cited on page(s) 104

[50] A. Hameed, A. Preese, and D. Sleeman. Ontology reconciliation. In S. Staab and R. Studer, editors, *Handbook on ontologies*, pages 231–250. Springer–Verlag, 2004. DOI: 10.1007/978-3-540-92673-3 Cited on page(s) 93

[51] Bernhard Haslhofer and Antoine Isaac. data.europeana.eu – the Europeana linked open data pilot. In *International Conference on Dublin Core and Metadata Applications (DC 2011)*, 2011. http://dcevents.dublincore.org/index.php/IntConf/dc-2011/paper/view/55/14. Cited on page(s) 43

[52] M. Hearst, A. Elliott, J. English, R. Sinha, K. Swearingen, and K.-P. Lee. Finding the flow in web site search. *CACM*, 45(9):42–49, 2002. Cited on page(s) 119

[53] T. Heath and C. Bizer. *Linked Data: Evolving the Web into a Global Data Space (1st edition)*. Synthesis Lectures on the Semantic Web: Theory and Technology. Morgan & Claypool, 2011. Cited on page(s) 8, 10, 33

[54] Dominik Heckmann, Tim Schwartz, Boris Brandherm, Michael Schmitz, and Margeritta von Wilamowitz-Moellendorff. Gumo – the general user model ontology. In Liliana Ardissono, Paul Brna, and Antonija Mitrovic, editors, *User Modeling*, pages 428–432. Springer–Verlag, 2005. Cited on page(s) 118

[55] Riikka Henriksson, Tomi Kauppinen, and Eero Hyvönen. Core geographical concepts: Case finnish geo-ontology. In *Location and the Web (LocWeb) 2008 workshop, 17th International World Wide Web Conference WWW 2008*, volume 300 of *ACM International Conference Proceeding Series*, pages 57–60, 2008. Cited on page(s) 71

[56] J. H. Herlocker, J. A. Konstan, and J. Riedl. Explaining collaborative filtering recommendations. In *Computer Supported Cooperative Work*, pages 241–250. ACM, 2000. Cited on page(s) 113

[57] Pascal Hitzler, Markus Krötzsch, and Sebastian Rudolph. *Foundations of Semantic Web technologies*. Springer–Verlag, 2010. Cited on page(s) 77, 81, 83, 86

[58] Johannes Hoffart, Fabian M. Suchanek, Klaus Berberich, and Gerhard Weikum. Yago: A spatially and temporally enhanced knowledge base from Wikipedia. *Artificial Intelligence*, 2012. Forth-coming. Cited on page(s) 70

[59] Aidan Hogan, Andreas Harth, Jürgen Umrich, and Stefan Decker. Towards a scalable search and query engine for the web. In *WWW '07: Proceedings of the 16th international conference on World Wide Web*, pages 1301–1302. Association of Computing Machinery, New York, 2007. Cited on page(s) 32

[60] M. Holi and E. Hyvönen. Fuzzy view-based semantic search. In *Proceedings of the 1st Asian Semantic Web Conference (ASWC2006), Beijing, China*. Springer–Verlag, 2006. Cited on page(s) 119

[61] Laura Hollink. *Semantic annotation for retrieval of visual resources*. PhD thesis, Free Univerity of Amsterdam, 2006. SIKS Dissertation Series, No. 2006-24. Cited on page(s) 108

[62] E. Hyvönen and E. Mäkelä. Semantic autocompletion. In *Proceedings of the first Asia Semantic Web Conference (ASWC 2006), Beijing*. Springer–Verlag, 2006. Cited on page(s) 111

[63] E. Hyvönen, E. Mäkela, M. Salminen, A. Valo, K. Viljanen, S. Saarela, M. Junnila, and S. Kettula. MuseumFinland—Finnish museums on the semantic web. *Journal of Web Semantics*, 3(2):224–241, 2005. Cited on page(s) 14, 20, 65, 86, 111

[64] E. Hyvönen, T. Ruotsalo, T. Häggströ, M. Salminen, M. Junnila, M. Virkkilä, M. Haaramo, T. Kauppinen, E. Mäkelä, and K. Viljanen. CultureSampo—Finnish culture on the semantic web. The vision and first results. In *Semantic Web at Work—Proceedings of STeP 2006*. Finnish AI Society, Espoo, Finland, 2006. Also in: Klaus Robering (Ed.), Information Technology for the Virtual Museum. LIT Verlag, 2008. Cited on page(s) 86, 114

[65] E. Hyvönen, S. Saarela, and K. Viljanen. Application of ontology techniques to view-based semantic search and browsing. In *The Semantic Web: Research and Applications. Proceedings of the First European Semantic Web Symposium (ESWS 2004)*. Springer–Verlag, 2004. Cited on page(s) 112, 119

[66] E. Hyvönen, M. Salminen, and M. Junnila. Annotation of heterogeneous database content for the semantic web. In *Proceedings of SemAnnot2004, Hiroshima, Japan*, Nov 2004. http://www.seco.tkk.fi/publications/2004/hyvonen-salminen-et-al-annotation-of-heterogeneous-2004.pdf. Cited on page(s) 100

[67] E. Hyvönen, K. Viljanen, E. Mäkelä, T. Kauppinen, T. Ruotsalo, O. Valkeapää, K. Seppälä, O. Suominen, O. Alm, R. Lindroos, T. Känsälä, R. Henriksson, M. Frosterus, J. Tuominen, R. Sinkkilä, and J. Kurki. Elements of a national semantic web infrastructure—case study Finland on the semantic web (invited paper). In *Proceedings of the First International Semantic*

Computing Conference (IEEE ICSC 2007), Irvine, California, pages 216–223. IEEE Press, Sept 2007. Cited on page(s) 10

[68] E. Hyvönen, K. Viljanen, J. Tuominen, and K. Seppälä. Building a national semantic web ontology and ontology service infrastructure—the FinnONTO approach. In *Proceedings of the 5th European Semantic Web Conference (ESWC 2008).* Springer–Verlag, 2008. Cited on page(s) 90, 94, 104

[69] Eero Hyvönen. Preventing interoperability problems instead of solving them. *Semantic Web – Interoperability, Usability, Applicability,* 1(1–2):33–37, 2010. Cited on page(s) 122

[70] Eero Hyvönen, Olli Alm, and Heini Kuittinen. Using an ontology of historical events in semantic portals for cultural heritage. In *Proceedings of the Cultural Heritage on the Semantic Web Workshop at the 6th International Semantic Web Conference (ISWC 2007),* 2007. http://www.cs.vu.nl/~laroyo/CH-SW.html. Cited on page(s) 75

[71] Eero Hyvönen, Thea Lindquist, Juha Törnroos, and Eetu Mäkelä. History on the semantic web as linked data – an event gazetteer and timeline for the World War I. In *Proceeedings of CIDOC 2012 – Enriching Cultural Heritage, Helsinki, Finland.* CIDOC, June 2012. http://www.cidoc2012.fi/en/cidoc2012/programme. Cited on page(s) 75

[72] Eero Hyvönen, Eetu Mäkelä, Tomi Kauppinen, Olli Alm, Jussi Kurki, Tuukka Ruotsalo, Katri Seppälä, Joeli Takala, Kimmo Puputti, Heini Kuittinen, Kim Viljanen, Jouni Tuominen, Tuomas Palonen, Matias Frosterus, Reetta Sinkkilä, Panu Paakkarinen, Joonas Laitio, and Katariina Nyberg. CultureSampo – Finnish culture on the Semantic Web 2.0. Thematic perspectives for the end-user. In *Museums and the Web 2009, Proceedings.* Archives and Museum Informatics, Toronto, 2009. Cited on page(s) 20, 33, 116

[73] Eero Hyvönen, Jouni Tuominen, Tomi Kauppinen, and Jari Väätäinen. Representing and utilizing changing historical places as an ontology time series. In Naveen Ashish and Amit Sheth, editors, *Geospatial Semantics and Semantic Web: Foundations, Algorithms, and Applications.* Springer–Verlag, 2011. Cited on page(s) 73, 95

[74] Dietmar Jannach, Markus Zanker, Alexander Felfernig, and Gerhard Friedrich. *Recommender Systems. An introduction.* Cambridge University Press, Cambridge, UK, 2011. Cited on page(s) 113

[75] K. Järvelin, J. Kekäläinen, and T. Niemi. ExpansionTool: Concept-based query expansion and construction. *Information Retrieval,* 4(3/4):231–255, 2001. Cited on page(s) 73

[76] M. Jensen. Vizualising complex semantic timelines. NewsBlip Research Papers, Report NBTR2003-001, 2003. http://www.newsblip.com/tr/. Cited on page(s) 75

[77] M. Junnila, E. Hyvönen, and M. Salminen. Describing and linking cultural semantic content by using situations and actions. In Klaus Robering, editor, *Information Technology for the Virtual Museum*. LIT Verlag, Berlin, 2008. Cited on page(s) 75, 113

[78] T. Känsälä and E. Hyvönen. A semantic view-based portal utilizing Learning Object Metadata. In *Semantic Web Applications and Tools Workshop, the 1st Asian Semantic Web Conference (ASWC2006)*, 2006. http://www.seco.hut.fi/publications/2006/kansala-hyvonen-2006-semantic-portal-lom.pdf. Cited on page(s) 118

[79] Akrivi Katifori, Constantis Halatsis, George Lepouras, Costas Vassilakis, and Eugeniua Giannopoulou. Ontology visualization methods – a survey. *ACM Computing Surveys*, 39(4), 2007. Article 10. Cited on page(s) 115, 119

[80] T. Kauppinen, C. Deichstetter, and E. Hyvönen. Temp-O-Map: Ontology-based search and visualization of spatio-temporal maps. In *Demo track at the European Semantic Web Conference ESWC 2007, Innsbruck, Austria*, June 4–5 2007. Cited on page(s) 116

[81] T. Kauppinen, R. Henriksson, J. Väätäinen, C. Deichstetter, and E. Hyvönen. Ontology-based modeling and visualization of cultural spatio-temporal knowledge. In *Semantic Web at Work—Proceedings of STeP 2006*. Finnish AI Society, Espoo, Finland, Nov 2006. Cited on page(s) 116

[82] Tomi Kauppinen, Glauco Mantegari, Panu Paakkarinen, Heini Kuittinen, Eero Hyvönen, and Stefania Bandini. Determining relevance of imprecise temporal intervals for cultural heritage information retrieval. *International Journal of Human-Computer Studies*, 86(9):549–560, September 2010. Cited on page(s) 74

[83] Tomi Kauppinen, Kimmo Puputti, Panu Paakkarinen, Heini Kuittinen, Jari Väätäinen, and Eero Hyvönen. Learning and visualizing cultural heritage connections between places on the semantic web. In *Proceedings of the Workshop on Inductive Reasoning and Machine Learning on the Semantic Web (IRMLeS2009), The 6th Annual European Semantic Web Conference (ESWC2009)*, 2009. Cited on page(s) 117

[84] Tomi Kauppinen, Jari Väätäinen, and Eero Hyvönen. Creating and using geospatial ontology time series in a semantic cultural heritage portal. In *Proceedings of the 5th European Semantic Web Conference (ESWC 2008)*, pages 110–123. Springer–Verlag, 2008. Cited on page(s) 95

[85] Suvi Kettula and Eero Hyvönen. Process-centric cataloguing of intangible cultural heritage. In *Proceeedings of CIDOC 2012 – Enriching Cultural Heritage, Helsinki, Finland*, 2012. http://www.cidoc2012.fi/en/cidoc2012/programme. Cited on page(s) 55, 113

[86] Barbara Ann Kipfer, editor. *Roget's International Thesaurus, 7th Edition*. Harper Collins Publishers, New York, 2011. Cited on page(s) 58

[87] R. Kishore, R. Ramesh, and R. Sharman, editors. *Ontologies: A Handbook of Principles, Concepts and Applications in Information Systems*. Springer–Verlag, 2006. Cited on page(s) 71

[88] Jussi Kurki and Eero Hyvönen. Authority control of people and organizations on the semantic web. In *Proceedings of the International Conferences on Digital Libraries and the Semantic Web 2009 (ICSD2009), Trento, Italy*, September 2009. http://www.seco.tkk.fi/publications/2009/kurki-hyvonen-onki-people-2009.pdf. Cited on page(s) 68

[89] Jussi Kurki and Eero Hyvönen. Collaborative metadata editor integrated with ontology services and faceted portals. In *Workshop on Ontology Repositories and Editors for the Semantic Web (ORES 2010), the Extended Semantic Web Conference ESWC 2010, Heraklion, Greece*, 2010. CEUR Workshop Proceedings, Vol 596, http://ceur-ws.org/. Cited on page(s) 104

[90] Michael S. Lew, Nicu Sebe, Chabane Djeraba, and Ramesh Jain. Content-based multimedia information retrieval: state of the art and challenges. *ACM Transactions on Multimedia computing, communications, and applications*, pages 1–19, Feb 2006. Cited on page(s) 108, 119

[91] Thea Lindquist, Eero Hyvönen, Juha Törnroos, and Eetu Mäkelä. Leveraging linked data to enhance subject access - a case study of the University of Colorado Boulder's World War I collection online. In *World Library and Information Congress: 78th IFLA General Conference and Assembly, Helsinki*. International Federation of Library Associations and Institutions (IFLA), 2012. http://conference.ifla.org/ifla78. Cited on page(s) 75

[92] John Lloyd. *Foundations of Logic Programming (2nd edition)*. Springer–Verlag, 1987. DOI: 10.1007/978-3-642-83189-8 Cited on page(s) 86

[93] Olivia Madison, John Byrum, Suzanne Jouguelet, Dorothy McGarry, Nancy Williamson, and Maria Witt, editors. *Functional requirements for bibliographic records. Final Report*. International Federation of Library Associations and Institutions (IFLA), 2009. http://www.ifla.org/files/cataloguing/frbr/frbr_2008.pdf. Cited on page(s) 46, 47

[94] A. Maedche, S. Staab, N. Stojanovic, R. Struder, and Y. Sure. Semantic portal—the SEAL approach. Technical report, Institute AIFB, University of Karlsruhe, Germany, 2001. Cited on page(s) 20

[95] E. Mäkelä, K. Viljanen, O. Alm, J. Tuominen, O. Valkeapää, T. Kauppinen, J. Kurki, R. Sinkkilä, T. Känsälä, R. Lindroos, O. Suominen, T. Ruotsalo, and E. Hyvönen. Enabling the semantic web with ready-to-use semantic widgets. In L. Nixon, R. Cuel, and C. Bergamini, editors, *First Industrial Results of Semantic Technologies, proceedings, co-located with ISWC 2007 + ASWC 2007, Busan, Korea*, 2007. CEUR Workshop Proceedings, Vol 293, http://ceur-ws.org. Cited on page(s) 119

[96] Eetu Mäkelä, Kaisa Hypén, and Eero Hyvönen. BookSampo—lessons learned in creating a semantic portal for fiction literature. In *Proceedings of ISWC-2011, Bonn, Germany*. Springer-Verlag, 2011. DOI: 10.1007/978-3-642-25093-4_12 Cited on page(s) 33

[97] Eetu Mäkelä, Kaisa Hypén, and Eero Hyvönen. Fiction literature as linked open data—the BookSampo dataset. *Semantic Web – Interoperability, Usability, Applicability*, 2012. Accepted for publication. Cited on page(s) 33

[98] Eetu Mäkelä, Aleksi Lindblad, Jari Väätäinen, Rami Alatalo, Osma Suominen, and Eero Hyvönen. Discovering places of interest through direct and indirect associations in heterogeneous sources—the TravelSampo system. In *Terra Cognita 2011: Foundations, Technologies and Applications of the Geospatial Web*, 2011. CEUR Workshop Proceedings, Vol 798, http://ceur-ws.org/. Cited on page(s) 33

[99] Eetu Mäkelä, Tuukka Ruotsalo, and Hyvönen. How to deal with massively heterogeneous cultural heritage data—lessons learned in CultureSampo. *Semantic Web – Interoperability, Usability, Applicability*, 3(1), 2012. DOI: 10.3233/SW-2012-0049 Cited on page(s) 20, 33

[100] Eetu Mäkelä, Osma Suominen, and Eero Hyvönen. Automatic exhibition generation based on semantic cultural content. In Aroyo et al. [6]. http://www.cs.vu.nl/~laroyo/CH-SW.html. Cited on page(s) 117

[101] Christopher D. Manning, Prabhakar Raghavan, and Hinrich Schütze. *Introduction to Information Retrieval*. Cambridge University Press, Cambridge, UK, 2008. DOI: 10.1017/CBO9780511809071 Cited on page(s) 119

[102] Christopher D. Manning and Hinrich Schütze. *Foundations of Statistical Natural Language Processing*. The MIT Press, Cambridge, Massachusetts, 1999. Cited on page(s) 110

[103] A. Maple. Faceted access: A review of the literature. Technical report, Working Group on Faceted Access to Music, Music Library Association, 1995. Cited on page(s) 119

[104] Suvodeep Mazumdar, Daniela Petrelli, and Fabio Ciravegna. Exploring user and system requirements of linked data visualization through a visual dashboard approach. *Semantic Web – Interoperability, Usability, Applicability*, 2012. In press. Cited on page(s) 117

[105] Willard McCarty. *Humanities Computing*. Palgrave, London, 2005. DOI: 10.1057/9780230504219 Cited on page(s) 11

[106] Pablo N. Mendes, Hannes Mühleise, and Christian Bizer. Sieve: Linked data quality assessment and fusion. In *EDBT/ICDT 2012 Joint Conference, Electronic Conference Proceedings, March 26–30, 2012, Berlin, Germany*, 2012. http://www.edbt.org/Proceedings/2012-Berlin/workshops_toc.html. DOI: 10.1145/2320765.2320803 Cited on page(s) 105

[107] Stuart E. Middleton, David De Roure, and Nigel R. Shadbolt. Ontology-based recommender systems. In Staab and Studer [139], pages 779–796. DOI: 10.1007/978-3-540-92673-3 Cited on page(s) 119

[108] Peter Mika and Mike Potter. Metadata statistics for a large web corpus. In *Proceedings of Linked Data on the Web Workshop (LDOW 2012), at the 21th International World Wide Web Conference (WWW 2012)*, 2009. Cited on page(s) 33

[109] Gabor Nagypal, Richard Deswarte, and Jan Oosthoek. Applying the semantic web: The VICODI experience in creating visual contextualization for history. *Lit Linguist Computing*, 20(3):327–349, 2005. DOI: 10.1093/llc/fqi037 Cited on page(s) 75

[110] N. F. Noy. Ontology mapping. In Staab and Studer [139], pages 573–590. DOI: 10.1007/978-3-540-92673-3 Cited on page(s) 93

[111] Natalya F. Noy, Nigam H. Shah, Patricia L. Whetzel, Benjamin Dai, Michael Dorf, Nicholas Griffith, Clement Jonquet, Daniel L. Rubin, Margaret-Anne Storey, Christopher G. Chute, and Mark A. Musen. BioPortal: ontologies and integrated data resources at the click of a mouse. *Nucleic Acids Research*, 37(Web Server issue):170–173, 2009. DOI: 10.1093/nar/gkp440 Cited on page(s) 77

[112] Natasha F. Noy and Mathieu d'Aquin. Where to publish and find ontologies? A survey of ontology libraries. *Web Semantics: Science, Services and Agents on the World Wide Web*, 11(0), 2011. DOI: 10.1016/j.websem.2011.08.005 Cited on page(s) 32, 77

[113] C. K. Ogden and I. A. Richards. *The meaning of meaning: A Study of the Influence of Language upon Thought and of the Science of Symbolism*. Magdalene College, University of Cambridge, 1923. Cited on page(s) 60

[114] J. Park and S. Hunting. *XML Topic Maps: Creating and Using Topic Maps for the Web*. Addison-Wesley, New York, 2003. Cited on page(s) 10

[115] Glenn E. Patton, editor. *Functional Requirements for Authority Data – A Conceptual Model*. K. G. Saur, München, 2009. DOI: 10.1515/9783598440397 Cited on page(s) 47

[116] Daniel V. Pitti. Encoded archival description. An introduction and overview. *D-Lib Magazine*, 5(11), 1999. http://www.dlib.org/dlib/november99/11pitti.html. DOI: 10.1045/november99-pitti Cited on page(s) 41

[117] A. S. Pollitt. The key role of classification and indexing in view-based searching. Technical report, University of Huddersfield, UK, 1998. http://www.ifla.org/IV/ifla63/63polst.pdf. Cited on page(s) 119

[118] Yves Raimond and Samer Abdallah. The event ontology, 2007. http://motools.sourceforge.net/event/event.html. Cited on page(s) 75

[119] E. Relph. *Place and placeness*. Pilon, London, U.K., 1976. Cited on page(s) 71

[120] D. Reynolds, P. Shabajee, and S. Cayzer. Semantic Information Portals. In *Proceedings of the 13th International World Wide Web Conference on Alternate track papers & posters*, New York, NY, USA, May 2004. ACM Press. DOI: 10.1145/1013367.1013440 Cited on page(s) 19, 20

[121] Bruce G. Robertson. Fawcett: A toolkit to begin an historical semantic web. *Digital Studies / Le champ numerique*, 1(2), 2009. Cited on page(s) 75

[122] Y. Rui, T. Huang, and S. Chang. Image retrieval: current techniques, promising directions and open issues. *Journal of Visual Communication and Image Representation*, 10(4):39–62, April 1999. DOI: 10.1006/jvci.1999.0413 Cited on page(s) 108

[123] Anisa Rula. DC proposal: Towards linked data assessment and linking temporal facts. In *Proceedings of the 10th International Semantic Web Conference (ISWC 2011)*, pages 110–123. Springer–Verlag, 2010. DOI: 10.1007/978-3-642-25093-4_27 Cited on page(s) 105

[124] T. Ruotsalo and E. Hyvönen. An event-based method for making heterogeneous metadata schemas and annotations semantically interoperable. In *Proceedings of ISWC 2007 + ASWC 2007, Busan, Korea*, pages 409–422. Springer–Verlag, 2007. DOI: 10.1007/978-3-540-76298-0_30 Cited on page(s) 75, 86

[125] T. Ruotsalo and E. Hyvönen. A method for determining ontology-based semantic relevance. In *Proceedings of the International Conference on Database and Expert Systems Applications DEXA 2007, Regensburg, Germany*. Springer–Verlag, 2007. DOI: 10.1007/978-3-540-74469-6_66 Cited on page(s) 113

[126] S. Russell and P. Norvig. *Artificial Intelligence. A Modern Approach.* Prentice-Hall, Upper Saddle River, New Jersey, 2010. Cited on page(s) 10

[127] G. M. Sacco. Dynamic taxonomies: guided interactive diagnostic assistance. In N. Wickra-masinghe, editor, *Encyclopedia of Healthcare Information Systems*. Idea Group, 2005. Cited on page(s) 119

[128] Satya S. Sahoo, Wolfgang Halb, Sebastian Hellmann, Kingsley Idehen, Ted Thibodeau, Sören Auer, Juan Sequeda, and Ahmed Ezzat. A survey of current approaches for mapping of relational databases to RDF, 2009. http://www.w3.org/2005/Incubator/rdb2rdf/RDB2RDF_SurveyReport.pdf. Cited on page(s) 104

[129] G. Salton, A. Wong, and C. S. Yang. A vector space model for automatic indexing. *Communications of the ACM*, 18(11):613–620, 1975. DOI: 10.1145/361219.361220 Cited on page(s) 108

[130] Ansgar Scherp, Carsten Saathoff, and Thomas Franz. Event-Model-F, 2010. http://www.uni-koblenz-landau.de/koblenz/fb4/AGStaab/Research/ontologies/events. Cited on page(s) 75

[131] A. Schoulltz, A. Matteni, R. Isele, C. Bizer, and C. Becker. LDIF – linked data integration framework. In *Proceedings of the 2nd International Workshop on Consuming Linked Data (COLD 2011)*, 2011. CEUR Workshop Proceedings, Vol 782, http://ceur-ws.org. Cited on page(s) 86, 104

[132] Guus Schreiber, Alia Amin, Mark van Assem, Viktor de Boer, Lynda Hardman, Michiel Hildebrand, Laura Hollink, Zhisheng Huang, Janneke van Kersen, Marco de Niet, Borys Omelayenko, Jacco van Ossenbruggen, Ronny Siebes, Jos Taekema, Jan Wielemaker, and Bob Wielinga. Multimedian e-culture demonstrator. In *Proceedings of the 5th International Semantic Web Conference (ISWC 2006)*, pages 951–958, 2006. DOI: 10.1007/11926078_70 Cited on page(s) 20, 65, 86, 89, 114

[133] Ryan Shaw. LODE: An ontology for linking open descriptions of events, 2010. http://linkedevents.org/ontology/. Cited on page(s) 75

[134] A. Sheth, B. Aleman-Meza, I. B. Arpinar, C. Bertram, Y. Warke, C. Ramakrishnan, C. Halaschek, K. Anyanwu, D. Avant, F. S. Arpinar, and K. Kochut. Semantic association identification and knowledge discovery for national security applications. *Journal of Database Management on Database Technology*, 16(1):33–53, Jan–March 2005. DOI: 10.4018/jdm.2005010103 Cited on page(s) 86, 114, 119

[135] S. Sirmakessis, editor. *Adaptive and Personalized Semantic Web*. Springer–Verlag, 2006. DOI: 10.1007/3-540-33279-0 Cited on page(s) 117

[136] H. Southall, R. Mostern, and M. L. Berman. On historical gazetteers. *International Journal of Humanities and Arts Computing*, 5:127–145, 2011. DOI: 10.3366/ijhac.2011.0028 Cited on page(s) 72

[137] J. Sowa. *Knowledge Representation. Logical, Philosophical, and Computational Foundations*. Brooks/Cole, 2000. Cited on page(s) 35, 58, 86

[138] S. Staab, J. Angele, S. Decker, M. Erdmann, A. Hotho, A. Maedche, H.-P. Schnurr, R. Studer, and Y. Sure. Semantic community web portals. In *Proceedings of the 9th International World Wide Web Conference*. Elsevier, Amsterdam, 2000. Cited on page(s) 19

[139] S. Staab and R. Studer, editors. *Handbook on ontologies (2nd Edition)*. Springer–Verlag, 2009. DOI: 10.1007/978-3-540-92673-3 Cited on page(s) 77, 104, 125, 126, 127, 133

[140] J. Suliman. Facilitating access: Empowering small museums. In *Museums and the Web 2007, Proceedings*. Archives and Museum Informatics, Toronto, 2007. http://www.museumsandtheweb.com/mw2007/papers/suliman/suliman.html. Cited on page(s) 14

[141] Osma Suominen and Eero Hyvönen. Improving the quality of SKOS vocabularies with Skosify. In *Proceedings of the 18th International Conference on Knowledge Engineering and Knowledge Management (EKAW 2012)*. Springer–Verlag, 2012. DOI: 10.1007/978-3-642-33876-2_34 Cited on page(s) 105

[142] Osma Suominen, Kim Viljanen, and Eero Hyvönen. User-centric faceted search for semantic portals. In *Proceedings of the 4th European Semantic Web Conference (ESWC 2007)*, pages 356–370. Springer–Verlag, 2007. DOI: 10.1007/978-3-540-72667-8_26 Cited on page(s) 119

[143] Heidi Suonuuti. *A Guide to Terminology*. Finnish Centre for Technical Terminology/ Nordterm, Helsinki, Finland, 2001. ISBN 952-9794-14-2. Cited on page(s) 60, 66

[144] A. Taylor. *Introduction to cataloging and classification*. Library and Information Science Text Series. Libraries Unlimited, 2006. Cited on page(s) 68, 69, 95

[145] B. Tillett. Authority control: State of the art and new perspectives. In *International Conference on Authority Control*. Haworth Press, Binghamton, NY, 2004. Cited on page(s) 68

[146] M. Tuffield, D. Millard, and N. Shadbolt. Ontological approaches to modelling narrative. In *2nd AKT DTA Symposium*, Jan. 2006. Cited on page(s) 75

[147] Jouni Tuominen, Nina Laurenne, and Eero Hyvönen. Biological names and taxonomies on the semantic web – managing the change in scientific conception. In *Proceedings of the 8th Extended Semantic Web Conference (ESWC 2011)*. Springer–Verlag, 2011. DOI: 10.1007/978-3-642-21064-8_18 Cited on page(s) 76

[148] M. van Assem, V. Malaise, A. Miles, and G. Schreiber. A method to convert thesauri to SKOS. In *Proceedings of the 3rd European Semantic Web Conference (ESWC 2006)*. Springer–Verlag, 2006. DOI: 10.1007/11762256_10 Cited on page(s) 89

[149] M. van Assem, M. R. Menken, G. Schreiber, J. Wielemaker, and B. Wielinga. A method for converting thesauri to RDF/OWL. In *Proceedings of 3rd International Semantic Web Conference (ISWC 2004), Hiroshima, Japan*. Springer–Verlag, 2004. DOI: 10.1007/978-3-540-30475-3_3 Cited on page(s) 89

[150] Mark van Assem. *Converting and Integrating Vocabularies for the Semantic Web*. PhD thesis, VU University, 2010. Cited on page(s) 89, 104

[151] Willem Robert van Hage, Véronique Malaisé, Roxane Segers, Laura Hollink, and Guus Schreiber. Design and use of the simple event model (SEM). *Web Semantics: Science, Services and Agents on the World Wide Web*, 9(2):128–136, 2011. DOI: 10.1016/j.websem.2011.03.003 Cited on page(s) 75

[152] J. van Ossenbruggen, A. Amin, L. Hardman, M. Hildebrand, M. van Assem, B. Omelayenko, G. Schreiber, A. Tordai, V. de Boer, B. Wielinga, J. Wielemaker, M. de Niet, J. Taekema, M.-F. van Orsouw, and A. Teesing. Searching and Annotating Virtual Heritage Collections with Semantic-Web Techniques. In *Proceedings of Museums and the Web 2007*, San Francisco, California, March 2007. Archives and Museum Informatics, Toronto. Cited on page(s) 20, 89

[153] K. Viljanen, T. Känsälä, E. Hyvönen, and E. Mäkelä. ONTODELLA—a projection and linking service for semantic web applications. In *Proceedings of the 17th International Conference on Database and Expert Systems Applications (DEXA 2006), Krakow, Poland*. IEEE, September 4–8 2006. DOI: 10.1109/DEXA.2006.105 Cited on page(s) 113

[154] K. Viljanen, J. Tuominen, T. Känsälä, and E. Hyvönen. Distributed semantic content creation and publication for cultural heritage legacy systems. In *Proceedings of the 2008 IEEE International Conference on Distibuted Human–Machine Systems, Athens, Greece*. IEEE Press, 2008. Cited on page(s) 77, 118

[155] Ubbo Visser. *Intelligent information integration for the Semantic Web*. Springer–Verlag, 2004. DOI: 10.1007/b100348 Cited on page(s) 71

[156] Manolis Wallace, Marios C. Angelides, and Phivos Mylonas, editors. *Advances in Semantic Media Adaptation and Personalization*. Springer–Verlag, 2006. Cited on page(s) 119

[157] W. Wong, W. Liu, and M. Bennamoun. Ontology learning from text: A look back and into the future. *ACM Computing Surveys*, 44(4):1–36, 2012. article 20. DOI: 10.1145/2333112.2333115 Cited on page(s) 104

[158] David Wood, editor. *Linking Government Data*. Springer–Verlag, 2011. DOI: 10.1007/978-1-4614-1767-5 Cited on page(s) 10

[159] G. P. Zarri. Semantic annotations and semantic web using nkrl (narrative knowledge representation language). In *Proceedings of the 5th International Conference on Enterprise Information Systems, Angers, France (ICEIS 2003)*, pages 387–394, 2003. Cited on page(s) 75

[160] Marcia Lei Zeng, Maja Zumer, and Athena Salaba, editors. *Functional Requirements for Subject Authority Data (FRSAD). A Conceptual Model*. International Federation of Library Associations and Institutions (IFLA), 2010. `http://www.ifla.org/files/classification-and-indexing/functional-requirements-for-subject-authority-data/frsad-model.pdf`. Cited on page(s) 48

Author's Biography

EERO HYVÖNEN

Eero Hyvönen[1] is a professor of semantic media technology at the Aalto University, Department of Media Technology, and an adjunct professor of computer science at the University of Helsinki, Department of Computer Science. He directs the Semantic Computing Research Group SeCo[2] specializing in Semantic Web technologies and applications. A major theme in his research during the last years has been the development of a semantic web content infrastructure on a national scale in Finland and its applications in areas such as Cultural Heritage. Eero Hyvönen has published over 300 articles, papers, and books. With his SeCo group, he has received several international and national awards, including the Semantic Web Challenge Award (in 2004 and 2008), World Summit Award (WSA) (2010), and Apps4Finland – Doing Good with Open Data (2010). He acts on the editorial boards of *Semantic Web – Interoperability, Usability, Applicability, Semantic Computing, and International Journal of Metadata, Semantics, and Ontologies,* and has co-chaired and acted on the program committees of tens of major international conferences and workshops, such as ESWC, ISWC, IJCAI, WWW, ICSC, etc.

[1]http://www.seco.tkk.fi/u/eahyvone/
[2]http://www.seco.tkk.fi/

Index

Printed in the United States
by Baker & Taylor Publisher Services